→INTRODUCING
CHAOS

ZIAUDDIN SARDAR & IWONA ABRAMS

This edition published in the UK
in 2008 by Icon Books Ltd.,
Omnibus Business Centre,
39–41 North Road, London N7 9DP
email: info@iconbooks.co.uk
www.introducingbooks.com

Sold in the UK, Europe, South Africa
and Asia by Faber and Faber Ltd.,
Bloomsbury House, 74–77 Great
Russell Street, London WC1B 3DA
or their agents

Distributed in the UK, Europe, South
Africa and Asia by TBS Ltd.,
TBS Distribution Centre,
Colchester Road, Frating Green,
Colchester CO7 7DW

Previously published in Australia
in 2008 by Allen & Unwin Pty. Ltd.,
PO Box 8500, 83 Alexander Street,
Crows Nest, NSW 2065

Previously published in the UK and
Australia in 1998 under the title
Chaos for Beginners and
in 1999 as *Introducing Chaos*

Reprinted 1998, 2003, 2004, 2009

This edition published in the USA
in 2008 by Totem Books
Inquiries to: Icon Books Ltd.,
Omnibus Business Centre,
39–41 North Road,
London N7 9DP, UK

Distributed to the trade in the USA by
National Book Network Inc.,
4501 Forbes Boulevard, Suite 200,
Lanham, Maryland 20706

Distributed in Canada by
Penguin Books Canada,
90 Eglinton Avenue East, Suite 700,
Toronto, Ontario M4P 2Y3

ISBN: 978-1-84831-013-1

Originating editor: Richard Appignanesi

Printed by Gutenberg Press, Malta

Yin, Yang and Chaos

Ancient Chinese thought recognized that chaos and order are related. In Chinese myth, the dragon represents the principle of order, yang, which emerges from chaos. In some Chinese creation stories, a ray of pure light, yin, emerges out of chaos and builds the sky. Yin and yang, the female and male principles, act to create the universe. But even after they have emerged from chaos, yin and yang still retain the qualities of chaos. Too much of either brings back chaos.

Ancient Chaos

Hesiod, a Greek of the 8th century B.C., wrote the **Theogony**, a cosmological poem which states that "first of all Chaos came to be", and then the Earth and everything stable. The ancient Greeks seem to have accepted that chaos precedes order, in other words, that order comes from disorder.

Chaos Theory

Chaos theory is a new and exciting field of scientific inquiry.

The phenomenon of chaos is an astounding and controversial discovery that most respectable scientists would have dismissed as fantasy just a decade or so ago.

Why is Chaos Exciting?

Chaos is exciting for all these reasons ...

It connects our everyday experiences to the laws of nature by revealing the subtle relationships between *simplicity* and *complexity* and between *orderliness* and *randomness*.

It presents a universe that is at once deterministic and obeys the fundamental physical laws, but is capable of disorder, complexity and unpredictability.

It shows that predictability is a rare phenomenon operating only within the constraints that science has filtered out from the rich diversity of our complex world.

It opens up the possibility of simplifying complicated phenomena.

It combines imaginative mathematics with the awesome processing power of modern computers.

It casts doubt on the traditional model-building procedures of science.

It shows that there are inherent limits to our understanding and predicting the future at all levels of complexity.

It is strikingly beautiful! Shakespeare had it right when he had Hamlet say in Act 1, scene 5 ...

There are more things in heaven and earth, Horatio, Than are dreamt of in your philosophy.

Hi! I'm Cordiallia Cauliflower. Just look at what chaos has done to me!

Where Does Chaos Come From?

Three major recent developments have made chaos a household word.

1. Breathtaking computing power that enables researchers to perform hundreds of millions of complicated calculations in matters of seconds.

2. The rise in computing power has been accompanied by a growing scientific interest in irregular phenomena such as …

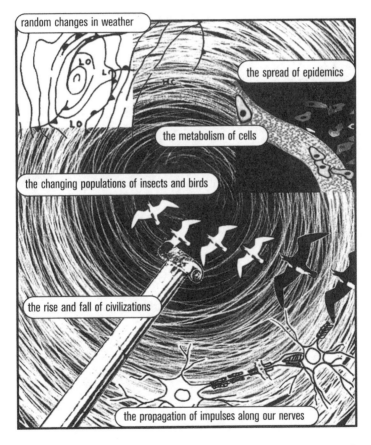

random changes in weather

the spread of epidemics

the metabolism of cells

the changing populations of insects and birds

the rise and fall of civilizations

the propagation of impulses along our nerves

3. Chaos theory was born when these developments were combined with the emergence of a new style of geometrical mathematics …

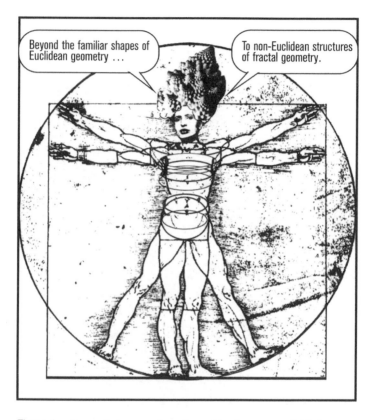

These developments have made an impact in almost every field of human endeavour. Chaos theory has been like a sea into which flow the rivers and tributaries of almost every discipline and subject – from mathematics, physics, astronomy, meteorology, biology, chemistry, medicine to economics and engineering, from the study of fluids and electrical circuits to the study of stock markets and civilizations.

Defining Chaos

Chaos has been variously defined. Here are just a few examples …

"A kind of order without periodicity."

"Apparently random recurrent behaviour in a simple deterministic (clock-work-like) system."

"The qualitative study of unstable aperiodic behaviour in deterministic nonlinear dynamical systems."

And here's another by a mathematician in the field, **Ian Stewart**.

Technical definitions of chaos are not easy to understand. So let's begin to familiarize ourselves with its terminology.

The Language of Chaos

Dynamic, Change and Variable

Chaos is a *dynamic* phenomenon. It occurs when something *changes*. Basically, there are two types of changes.

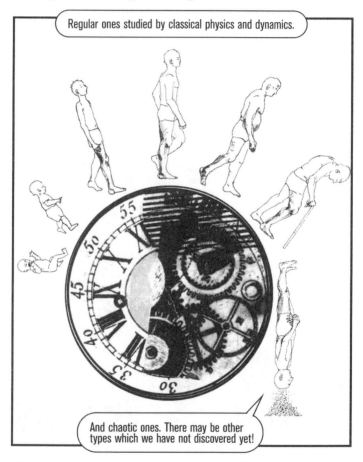

Regular ones studied by classical physics and dynamics.

And chaotic ones. There may be other types which we have not discovered yet!

What is changeable in a given situation is referred to as a **variable**.

Systems

Any entity that changes with time is called a **system**. Systems thus have variables. Here are some examples of systems.

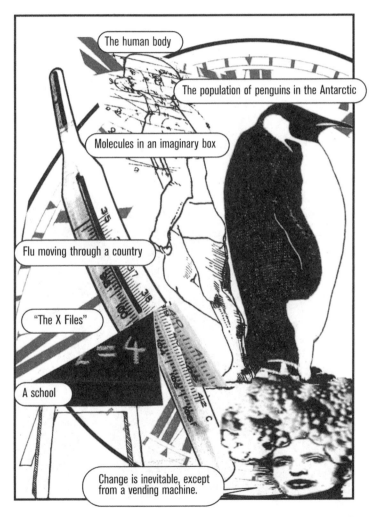

Defining Systems

A **deterministic system** is one that is predictable, stable and completely knowable. The classic example of a deterministic system is an old-fashioned grandfather clock. The balls on a snooker table behave within the boundaries of a deterministic system.

In classical physics, the universe itself was considered to be a deterministic system.

Give me the past and present co-ordinates of any system and I will tell you its future.

Pierre Simon Laplace (1749-1827), French mathematician.

In **linear systems**, variables are simply and directly related. Mathematically, a linear relationship can be expressed as a simple equation where the variables involved appear only to the power of one:

$x = 2y + z$

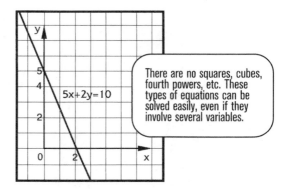

There are no squares, cubes, fourth powers, etc. These types of equations can be solved easily, even if they involve several variables.

Nonlinear relationships involve powers other than one. Here is a nonlinear equation:

$A = 3B^2 + 4C^3$

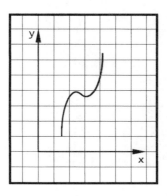

Such equations are much harder to analyze and frequently need the help of a computer to understand.

Periodic and Aperiodic Equations

A **period** is an interval of time characterized by the occurrence of a certain condition or event. A variable in a **periodic system** exactly repeats its past behaviour after the passage of a fixed interval of time – think of a swinging pendulum.

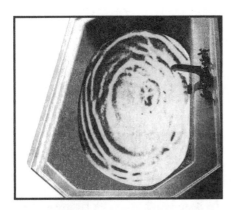

Aperiodic behaviour occurs when no variable affecting the state of the system undergoes a completely regular repetition of values – visualize the flow of water as it goes down a sink.

Unstable aperiodic behaviour is highly complex. It never repeats itself and continues to show the effects of any small perturbation to the system. This makes exact predictions impossible and produces a series of measurements that appear random.

That's why, in spite of our satellite observations and computer models, it is still impossible to predict the weather accurately.

What is Unstable Aperiodic Behaviour?

Behaviour that is unstable yet periodic is difficult to imagine – indeed, it appears to be a contradiction in terms. However, human history provides us with several examples of just such a phenomenon. It is possible to chart broad patterns in the rise and fall of civilizations. We can see that these patterns are *periodic*. But we know that events never actually repeat themselves exactly. In this realistic sense, history is *aperiodic*. We can also read in history textbooks that seemingly small unimportant events have led to long-lasting changes in the course of human affairs.

Until quite recently, our principal image of behaviour that is so complex as to be unstable and aperiodic was the image of a crowd.

Now that our perception has changed, we see such behaviour in even the commonest events: water dripping from a tap, a flag waving in the breeze, the fluctuation of animal populations.

Linear Systems

So: simply put, chaos is the occurrence of aperiodic, apparently random events in a deterministic system. In chaos there is order, and in order there lies chaos. The two are more closely connected than we ever thought before.

But since deterministic systems are predictable and stable, this seems to be illogical. As a matter of habit, humans have looked for patterns and linear relations in what they see.

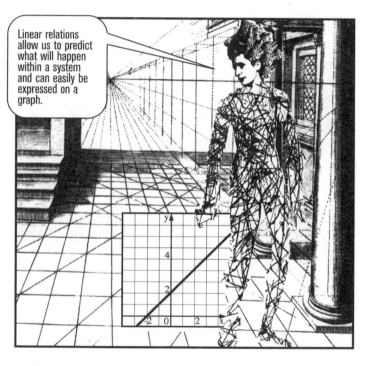

Linear relations allow us to predict what will happen within a system and can easily be expressed on a graph.

In other words, they form a straight line on the graph and we know where that line is going.

Linear relationships and equations are solvable. That makes them easy to think about and work with.

Nonlinear Complication

Nonlinear equations, on the other hand, cannot be solved. Friction, for example, often makes things difficult by introducing nonlinearity. Without friction, the amount of energy required to accelerate an object is expressed in a linear equation ...

force = mass x acceleration

Friction complicates things because the amount of energy changes, depending on how fast the object is moving.

Nonlinearity, therefore, changes the deterministic rules within a system and makes it difficult to predict what is going to happen.

There is a famous example of a nonlinear relationship in the history of chaos. **Robert May**, a biologist, was studying an imaginary population of fish. The mathematical model he used for the fish population was the equation $x_{next} = rx(1-x)$, where x represents the present population of fish in an area. When the parameter, **r** (rate of growth) was 2.7, he found the population to be .6292.

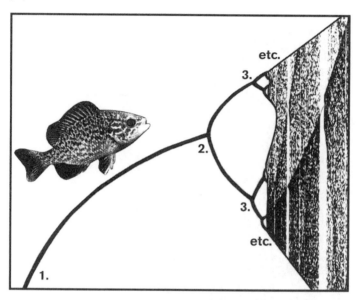

1. As the parameter rose, the final population rose slightly too, making a line that rose as it moved from left to right on the graph.

2. Suddenly, as the parameter passed 3, the line broke in two and May had to plot for two populations. This split meant that the population was going from a one-year cycle to a two-year cycle.

3. As the parameter rose further, the number of points doubled again and again. The behaviour was complex yet regular. Beyond a certain point, the graph became totally chaotic – and the graph was completely blacked in. Yet even in the midst of the chaos, stable cycles returned as the parameter was increased.

Most forces in real life are nonlinear. So why have we not discovered this before? The reason that chaotic behaviour has not been studied until now is because scientists reduced difficult nonlinear problems to simpler linear ones in order to analyze them. **Galileo**'s work with gravity provides us with a good example. Galileo (1564-1642), an Italian physicist, disregarded small nonlinearities in order to get neat results.

Feathers do not fall with the same speed as a ball, due to air resistance. H'm, so what ...

An ideal scientific world was created where regularities were isolated from actual experience and "disorder".

Since the advent of "modern" Western science, we have been living in a world which acts as if the platypus was the only animal in existence!

Feedback

Feedback, like nonlinearity, is also common in real-life events. Feedback is a characteristic of any system in which the output, or result, affects the input of the system, thus altering its operation.

Feedback is most commonly observed when a microphone is in use. Some of the output signal is literally "fed back" into the system and causes the screeching sounds that engineers and musicians dread. Feedback can, however, be useful in the production of amplifiers where it is deliberately looped back into a system.

Feedback is also observed on the trading floor and is actually a form of self-regulation.

We can also see feedback loops when an enzyme produces a copy of itself in a chemical reaction. This is a positive feedback loop. This happens when DNA becomes a living organism and is very common in organic chemistry.

However, scientists have tended to ignore feedback to create simple models that are easier to study and work with. They knew about feedback and complexities but did not understand them. For example, it is much easier to study population as a simple linear system than one involving feedback and complexity.

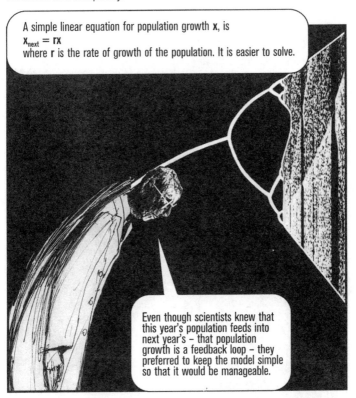

A simple linear equation for population growth **x**, is
$$x_{next} = rx$$
where **r** is the rate of growth of the population. It is easier to solve.

Even though scientists knew that this year's population feeds into next year's – that population growth is a feedback loop – they preferred to keep the model simple so that it would be manageable.

Oscillating systems become chaotic because they possess an element of feedback. Chaotic behaviour results when nonlinear forces are turned back on themselves. This is called *nonlinear feedback* – and is an essential prerequisite for chaos. Let's now have an example of nonlinear feedback.

The Three Body Problem

An example of a simple linear system that exhibits nonlinear feedback effect is the classic "three body problem" of gravitation. Consider a moon orbiting a planet. The path that the moon takes is well-known – it was fully described by **Sir Isaac Newton**'s (1642-1727) mathematical laws of gravity.

> But suppose we introduce a second moon of the same size as the first. Would the moons' orbits now be only slightly more difficult to calculate?

It turns out that the simple deterministic equations which govern this three-body system are "unsolvable". They cannot predict the long-term path of the orbiting moons.

The reason why the three-body problem cannot be solved is that gravity is a nonlinear force (specifically, it is "inverse square"), and in a three-body system each body exerts its force on the other two. This produces non-linear feedback and results in chaotic motion of the moons' orbits. But we have now "solved" the three-body problem by demonstrating that the orbits are inherently unpredictable. Such a solution would have been considered nothing short of sacrilege a few years ago.

An amateur Biblical scholar, **Immanuel Velikovsky** (1895-1979), was dismissed by astronomers as a complete crank when he argued in his *Worlds in Collision* (1948) that the orbits of Mars and Venus had changed drastically around 1000 B.C. His theory did help to resolve some difficulties with the chronology of the ancient world.

Chaos Modelling

During the past two decades, scientists working in fields as disparate as weather forecasting, fluid mechanics, chemistry and population biology have been developing models for natural phenomena that take nonlinearity and feedback into account. These models display two incongruous features. First, they consist of only a few simple equations. And second, solutions to these equations are complex and sometimes unpredictable. The analysis of these models, and similar behaviour in experiments, is what we now know as "chaos theory".

If we take the simple equation $x^2 + c = result$, where x is a complex number that changes and c is a fixed complex number, and continuously feed back the result into the changing number (x) slot – that is, we *iterate* the equation – chaotic patterns like these are produced …

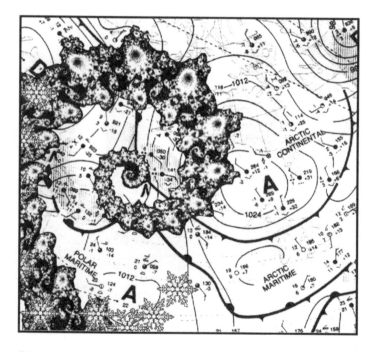

Questions of Long-Term Behaviour

Chaos theory works by asking questions about the long-term behaviour of a system. Instead of making predictions about the future state of a system, chaos attempts a qualitative study of the system by concentrating on behaviour that is unstable and aperiodic. Conventional astronomy, for example, is interested in knowing when a system of three planets will line up.

In contrast, chaos theory asks: what circumstances would lead to elliptical orbits?

Or circular ones?

What characteristics will **all** solutions of this system exhibit?

How does this system change from exhibiting one form of behaviour to another?

The Signature of Chaos

A distinguishing feature of systems studied by chaos theory is that unstable aperiodic behaviour can be found in mathematically simple systems. Very simple, rigorously defined mathematical models can display behaviour that is awesomely complex.

Another distinguishing characteristic of chaotic systems is their sensitive dependence on initial conditions – infinitesimally small changes at the start lead to bigger changes later. This behaviour is described as the *signature* of chaos.

This feature of chaotic systems is seen by some scientists as an important source of novelty and diversity in our natural world.

Other scientists see this as a boundary where human knowledge runs up against limitations – where nature decrees: "Here you can go no further."

The Little Devil

To explain sensitive dependence, mathematical physicist **David Ruelle** tells this story. "The little devil, presumably having nothing else to do, decides one day to upset your life. The devil does this by altering the motion of a single electron in the atmosphere. But you don't notice. Not yet. After a minute, the structure of turbulence in the air has changed. You still don't notice that anything is amiss. But after a couple of weeks, the change has taken on much larger proportions, and while you are having a picnic lunch with someone rather important, the skies open and a hailstorm begins.

Now you notice what the little devil has achieved. Actually, she wanted to kill you in a plane crash but I talked her out of it."

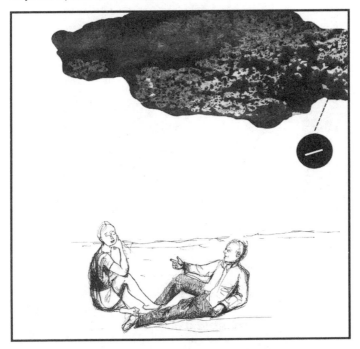

Let's now look at the history of chaos theory and meet the people who helped shape it.

Benoit Mandelbrot and Fractal Geometry

Benoit Mandelbrot (b. 1924), a Polish-born French mathematical physicist who worked for IBM, developed the field of **fractal geometry** which played a key role in the emergence of chaos theory. He did most of his pioneering work in the 1970s and published his findings in an illustrated and erudite book called *Fractals: Forms, Chance and Dimensions*. No one understood what he was on about – largely because the prose was difficult to fathom. In 1977, a much-refined version of that book was republished as *The Fractal Geometry of Nature* – and fractal geometry caught the imagination of scientists.

Chaos and Order in Economics

Mandelbrot, "a mathematical jack-of-all-trades", started his work in economics. Economists believed that small transient changes had nothing in common with large, long-term changes. Mandelbrot investigated this – but did not separate small changes from large ones. He looked at the system as a whole.

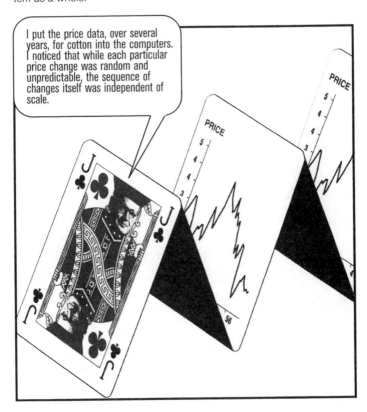

Indeed, curves for daily and monthly price changes were perfectly matched. The degree of variation had remained constant over sixty years, spanning two world wars and a depression. In other words: within the chaos there was order.

Chaos on the Telephone Lines

Mandelbrot also worked on telephone lines used to transmit information from computer to computer. Engineers were puzzled by the problem of noise in the lines. The current carries the information in "discrete packets" in the lines. But some spontaneous noise could not be eliminated. Sometimes it would wipe out the signal, creating an error. The interference was random, yet it occurred in clusters.

He found an hour with no errors. But when he divided the hour with errors into two, he found again, a period with no errors and a period with errors. Again, when he divided the period with errors into two, the same thing happened – one period error-free and one period with errors.

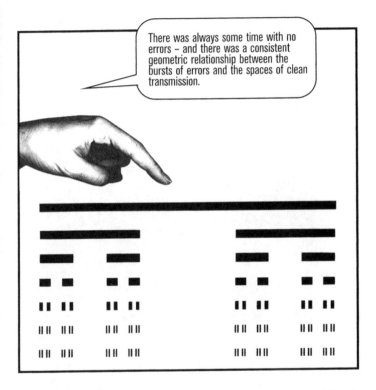

This phenomenon was incomprehensible to engineers, but mathematicians knew this as the **Cantor Set** – a pattern that is created by removing segments of a line, then removing segments of the segments through to infinity, leaving a dust of points arranged in clusters. Instead of increasing signal strength to drown out the interference, the engineers were advised to arrange a more modest signal, and to accept that errors would happen. They had to arrange a way to catch and modify them.

Measuring the Coast

In a well-known paper, Mandelbrot asked "How Long is the Coast of Britain?". Suppose we measure the British coastline with a metre stick. The answer will be approximate, as the little nooks and crannies will be overlooked by this particular measure.

But say we measure the coastline with a smaller scale, say 10 cm, and repeat the process. Here a greater length will be arrived at because the measure can go into those smaller spaces and count them.

If we set the divider at 5 cm, the measurement will be even bigger. So, if we measure the coastline with smaller and smaller scale, the answer will become larger and larger. As we approach towards a very small scale of measurement, the coastline becomes longer and longer without a limit.

Fractal Dimensionality

Mandelbrot suggested that what we observe depends on where we are positioned and how we measure it. Consider a football. From far away it looks like a two-dimensional disc. When we get closer, it becomes a three-dimensional object.

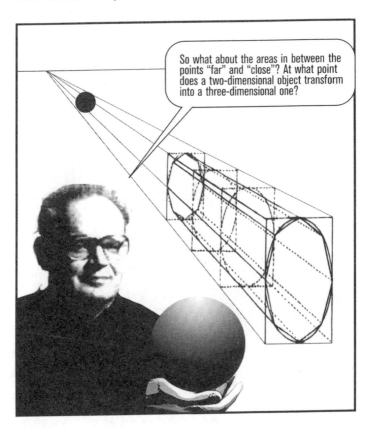

> So what about the areas in between the points "far" and "close"? At what point does a two-dimensional object transform into a three-dimensional one?

Mandelbrot described systems with fractional dimensionality with the term **fractals**. The coastline of Britain is an example of fractals. And, he argued, the only way to solve this problem is to move from ordinary three dimensions into what he called "fractal dimensions".

What are Fractals?

The geometry that we are all used to is attributed to **Euclid**, a Greek mathematician (*circa* 300 B.C.). Euclidean shapes are regular – triangles, squares, circles, rectangles. Fractal geometry is the geometry of special types of irregular shapes. Fractals are a way of measuring qualities that otherwise have no clear definition: the degree of roughness or brokenness or irregularity in an object.

Indeed, a fractal is a way of seeing infinity.

Mandelbrot: "I coined the word fractal in 1975 from the Latin *fractus*, which describes a broken stone – broken up and irregular. Fractals are geometrical shapes that, contrary to those of Euclid, are not regular at all. First, they are irregular all over. Secondly, they have the same degree of irregularity on all scales. A fractal object looks the same when examined from far away or nearby – it is self-similar."

Self-similarity implies that any subsystem of a fractal system is equivalent to the whole system. In the fractal triangle, each small triangle is structurally identical to the large one. Some fractals, though, are only *statistically* self-similar – their magnified small pieces do not superimpose on the entire system – but they do have the same general type of appearance.

Fractals can expose the abstract geometrical nature of chaos, particularly in the form of computer graphics.

Within the overall shape, there lies a repetitive pattern whose exquisite substructure characterizes the nature of chaos, indicating when predictability breaks down.

Fractals are Everywhere ...

Fractals also provide us with an immediate link with nature. Trees and mountains are examples of fractals. They are everywhere.

The Julia Set

Fractals can lead to beautiful graphics, and some fractals have been known for years. **Gaston Julia** and **Pierre Fatou**, during the First World War, discovered the **Julia set**. It explores imaginary numbers in a complex plane. Imaginary numbers are produced when we look for the square root of a negative number. The square root of -1 is considered to be i and the square root of -4 is $2i$. But at that time, nobody realized the significance of these sets for "physics of the real world".

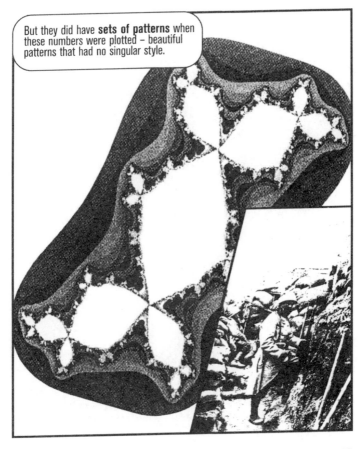

But they did have **sets of patterns** when these numbers were plotted – beautiful patterns that had no singular style.

The Use of Fractals

Nowadays, fractal geometry is used to describe many complex phenomena. Fractals help us understand turbulence, not just how it arises, but the motion of the turbulence itself.

Blood vessels can also be considered as fractals, as they can be divided down into smaller and smaller sections. They perform what has been described as "dimensional magic", squeezing a large surface area into a limited volume.

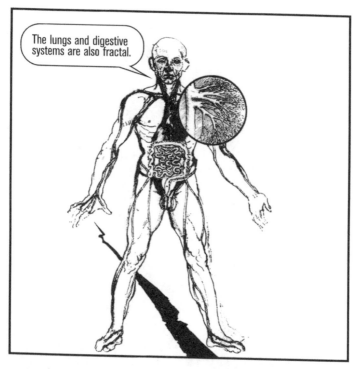

So are earthquakes. The distribution of earthquakes was known to fit a mathematical pattern. This pattern was picked up by geologists and found to be fractal. The fractal dimensions of a metal's surface also tell us a lot about its strength.

Mandelbrot has also lent his name to this famous fractal – it is known as (what else?) the **Mandelbrot set**.

Millions of people around the world watched fractal mathematics – without knowing it – when they sat through the *Star Wars* film trilogy. Images of alien landscapes in the films were generated on computer using fractals. Indeed, fractals are now an important part of special effects in films.

Edward Lorenz

Edward Lorenz (b. 1917), a meteorologist, was the first to record a known instance of chaotic behaviour. Lorenz started his postdoctoral work in 1948 at the Department of Meteorology, Massachusetts Institute of Technology. In 1955, he became the director of a project in statistical weather forecasting, a field that was pioneered by his department.

Following the example of astronomers in the 18th and 19th centuries, Lorenz used hand computation to estimate the solutions.

Later, using computer models of the earth's atmosphere and oceans, Lorenz studied the interrelationship between three nonlinear meteorological factors: temperature, pressure and wind speed.

I found that very small changes in the initial conditions produced widely varying and unpredictable responses. How could a simple three-equation model give such a bizarre result?

Lorenz was forced to conclude that this type of response was inherent in his model. In 1963, he published his results in a paper entitled "Deterministic Nonperiodic Flow" in the *Journal of the Atmospheric Sciences*. It took almost a decade for researchers to grasp the significance of this paper.

Small Differences, Big Consequences

Lorenz's discovery of the phenomena of chaos is often related in an interesting story. One day in 1961, the story goes, Lorenz decided to take a short cut with his weather machine. He wanted to examine one sequence at greater length. So instead of starting the whole computer run from the beginning, he started half-way through. He tapped in the numbers straight from an earlier printout and went away to get a coffee. When he came back he could hardly believe his eyes.

The newly generated weather was nowhere near the original. They were two completely different systems!

Then he realized what had happened. He had tapped in the number .506, the number stored on the printout, whereas the original number from the computer's memory was .506127. The small difference – one part in five thousand – was not inconsequential. Lorenz realized that minute differences in the initial conditions – like a puff of wind – could prove catastrophic.

The consequences of his discovery were spelled out by Lorenz in these words: "It implies that two states differing by imperceptible amounts may eventually evolve into two considerably different states. If, then, there is any error whatever in observing the present state – and in any real system such errors seem inevitable – an acceptable prediction of the instantaneous state in the distant future may well be impossible."

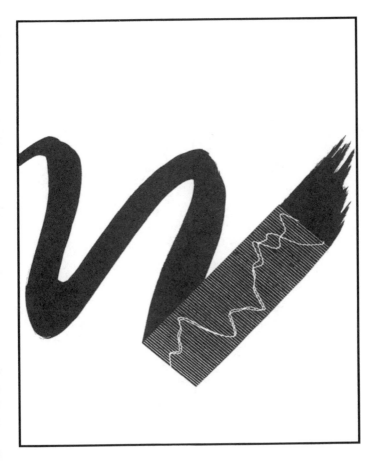

The Water Wheel Example

One example used by Lorenz to demonstrate chaos is the water wheel. This simple mechanical device is capable of astonishingly complicated behaviour.

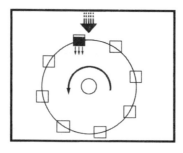

At certain slow speeds, the system works fine.

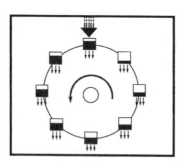

But when the water flow is increased, the wheel turns faster, the buckets have little time to fill up or become empty, and the behaviour of the system becomes chaotic.

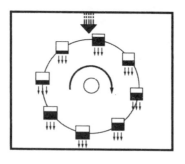

The spin will then slow down or even reverse. It never repeats itself in any predictable patterns under these conditions.

When the chaotic behaviour of the water wheel is plotted, it produces a very beautiful graph – a double spiral in space that is known as a "strange attractor".

Strange Attractors

Complex systems, in general, exhibit a property that mathematicians call **attractors**. Attractors represent the states to which the system eventually settles, depending on the properties of the system.

Imagine a marble swirling around a bowl. The marble eventually settles at the bottom of the bowl. The point at which the marble settles **attracts** the marble.

Another way of thinking about attractors is to look at some real-world situations where certain conceivable modes of behaviour just do not occur. A pendulum in a clock in good working order does not swing gently at times and violently at others. Arctic temperatures do not occur at the equator. Pigs do not normally fly. Unusual things that *do* occur thus belong to a special area – or to put it technically, a restricted set. This is the *set of attractors*.

Cultural and Identity Attractors

The cultural equivalent of attractors would be chiefs, tribes, states and what gives us identity, like religion, class and worldviews.

Chaotic Attractors

Now, there is a class of attractors that are a bit out of the ordinary – they are known as "chaotic" or "**strange attractors**".

They consist of infinite numbers of curves, surfaces or higher dimensional manifolds. They are, in fact, **fractal objects**.

The strange attractors live in a mathematical construct known as **phase space**. Phase space is an imaginary space – it is a way in which numbers can be turned into pictures, making a flexible map of all the information available. Let's define "phase space".

Representing Phase Space

We are familiar with architectural drawings that represent a three-dimensional building in a two-dimensional plane. But suppose instead of a fixed object (a building), we had a moving object – say, a pendulum. We can represent the horizontal and vertical motion of the pendulum in a two-dimensional graph.

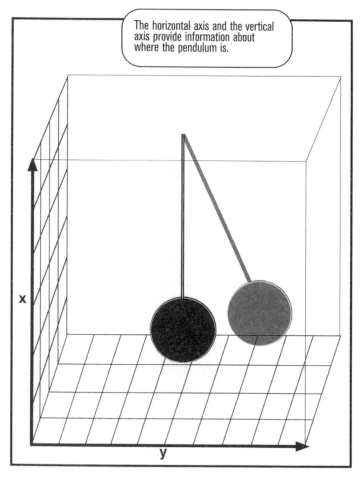

Similarly, phase space represents the state of an object in a multi-dimensional plane. The motion of the simple pendulum could be shown on a graph where the x-axis is the angle of displacement from the vertical and the y-axis is the angular velocity. On this phase space diagram, the simple pendulum shows as a circle.

Phase space turns dull statistical data into a telling picture, abstracting all the essential information from the moving parts and providing us with an easy to grasp overview of the system's behaviour over time.

In phase space, the complete state of knowledge about a dynamical system at a single instant in time collapses to a point. That point then is the dynamical system at that instant. At the next instant, the system will have changed, and so the point moves.

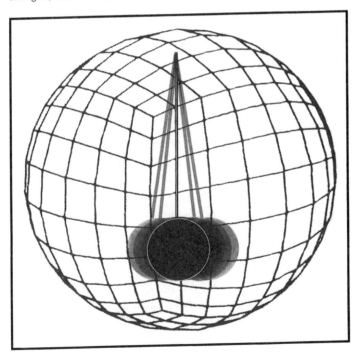

Phase space makes a dynamic system easy to watch.

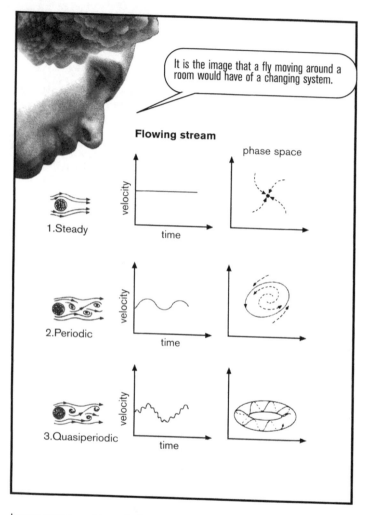

It is the image that a fly moving around a room would have of a changing system.

Flowing stream

phase space

1. Steady

velocity / time

2. Periodic

velocity / time

3. Quasiperiodic

velocity / time

Loops correspond to periodicity, twists correspond to change, and empty space corresponds to physical impossibility.

What is strange about strange attractors? First: they look strange. A multi-dimensional imaginary object is bound to look strange. Second: the motion on the strange attractors has sensitive dependence on initial conditions. Third: strange attractors reconcile contradictory effects: (a) they are attractors, which means that nearby trajectories *converge* on them;

and (b) they exhibit sensitive dependence on initial conditions, which means that trajectories initially close together on the attractors *diverge* rapidly.

Fourth: and this is the tricky bit – while strange attractors exist in an infinite dimensional space (the phase space), they themselves have only *finite* dimensions.

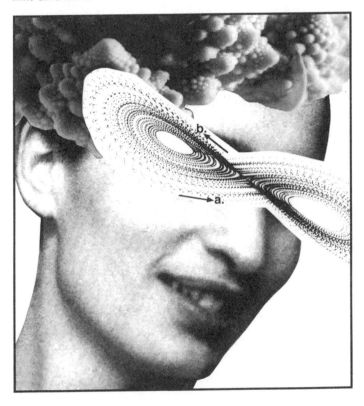

The Lorenz Attractor

The most famous strange attractor is known as the Lorenz attractor because it was first discovered by Lorenz. This is what it looks like.

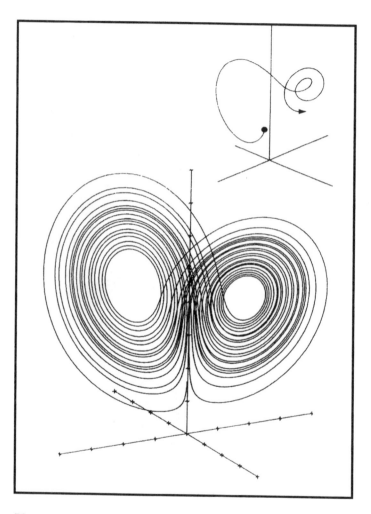

The term "strange attractor" was coined by David Ruelle, Professor of Theoretical Physics at the Institut des Hautes Études Scientifiques, Bures-sur-Yvette, France. He introduced the term in the early 1970s in a paper, written with a colleague, in which he proposed that fluid turbulence is an example of chaos.

There have been objections to the term "strange attractor". For example, the Russian mathematicians **Boris Chirikov** and **Felix Izrailve** suggest that strange attractors look strange only to a stranger.

However, the name is too attractive for most scientists, and the term has stuck. The strange attractors have fuelled the fire of chaos theory. Researchers now look everywhere for strange attractors – in any system that appears to be acting randomly.

The Butterfly Effect

Lorenz is also associated with the idea of "the Butterfly effect". In 1972, he presented a paper at a conference in Washington, entitled "Does the Flap of a Butterfly's Wings in Brazil Set Off a Tornado in Texas?". He did not actually answer the question.

54

Two factors ensured that "the Butterfly effect" would become an emblem of chaos. First, amongst the earliest chaotic systems studied by Lorenz was the famous "strange attractor" that resembled a butterfly. It was natural for some people to assume that "the Butterfly effect" was named after this attractor. Second, "the Butterfly effect" was given a mythical status by **James Gleick** in his best-selling book, *Chaos* (1988).

David Ruelle

Mathematical physicist David Ruelle gave chaos theory a kick-start with his work on turbulence. For decades, turbulence had been a major problem for physicists. **Werner Heisenberg** (1901-76), who contributed the "Uncertainty Principle" to quantum physics, probably still worried about it even on his deathbed.

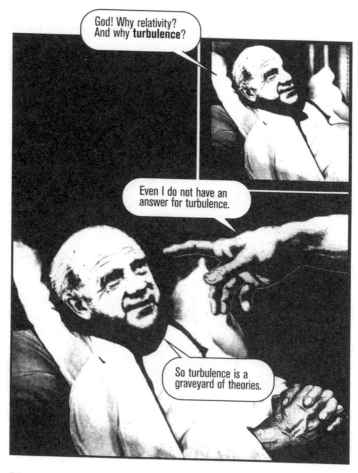

What is Turbulence?

You can see turbulence in action by making a quick visit to the bathroom. Open the tap by a very small amount, gently, and you can arrange a steady stream of water between the tap and the sink. The column of water seems as if motionless – but, of course, the tap is running.

If you open the tap a little more – carefully – you may be able to arrange regular pulsations of the water column. This is **periodic motion**.

When the tap is open a little further, the pulsations become irregular. Finally, when the tap is wide open, you have a total mess – very irregular flow. This is **turbulence**.

How Does Turbulence Happen?

Turbulence is a mess of disorder at all scales. It is unstable and highly dissipative, meaning that it drains energy and creates drag. The puzzle is *how* does a smooth, stable flow *become* turbulent?

Ruelle's Approach

The equations of fluid flow are, in fact, mostly impossible to solve. They are nonlinear partial differential equations. Ruelle decided he would try to work out an abstract alternative to the usual approach.

x describes a fluid in turbulent motion as quasi-periodic functions of time

$$x(t) = f(\omega_1 t, ..., \omega_k t)$$

I suggested that three independent motions cause all the complexities of turbulence.

Ruelle published his analysis in 1971, in a paper entitled "On the Nature of Turbulence", co-written with **Floris Takens**, a Dutch mathematician. (Actually, Ruelle was an editor of the journal, and he accepted the paper himself for publication. This is not a recommended procedure in general, but he felt it was justified in this particular case.)

Even though much of the mathematics in Ruelle's paper was obscure or simply wrong, it contained elements that made a lasting impression.

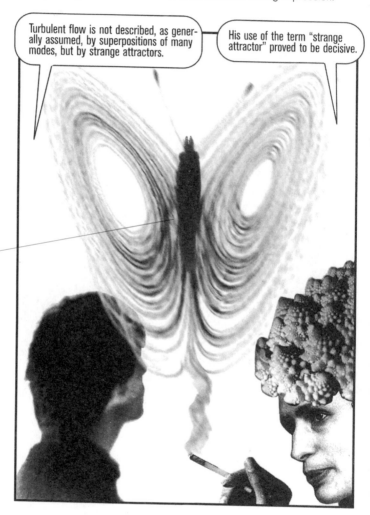

Turbulent flow is not described, as generally assumed, by superpositions of many modes, but by strange attractors.

His use of the term "strange attractor" proved to be decisive.

New questions arose. How could an infinite number of loops and spirals be contained within finite space? How can so much be going on in such a tiny space? Why infinite logic to understand what a point in time is going to do? Ruelle suspected that the visible patterns in turbulent flow – that come and go at random – must be related to some laws that had not yet been discovered. One thing that was known about turbulence was that a broad spectrum of cycles was present at once. But how could this be represented? Could it arise from simple equations? The attractor would have to be stable and represent the final state of the dynamic system. It would also have to be non-periodic, never repeating itself and never cutting across itself.

To produce **every** rhythm, it would have to be infinitely long within a finite space – a fractal.

But the term "strange attractor" was not then known. Ruelle argued that such a thing had to exist.

He was vindicated when strange attractors began to appear as far apart as Germany and Japan.

Robert May and Animal Populations

Robert May (b. 1936), an Australian mathematical biologist who worked at Princeton University and later became Royal Society Research Professor at Oxford University, was responsible for pioneering work in population dynamics that helped shape chaos theory.

I studied **predator-prey** populations and found that nonlinear feedback forces in the environment produced pseudo-random changes in animal population.

The population of a particular species, say antelopes, varies from year to year and the total number in a particular year provides a good indication for numbers during the next.

Unlike the pendulum or a ball on a snooker table, animal populations do not have my "Newton's law".

The conventional wisdom said that usually the population would fluctuate around a point – with predators, food supply, the environment and disease all keeping the numbers in check.

Thus, if a population rose above a particular level, food supplies would dwindle, more of the animals would starve and die, and then the population would return to its "normal" state. The year with the largest population would therefore follow a year with medium-sized population.

May's Bifurcations

In the 1970s, May's research revealed that the equations used to describe fluctuations in animal populations were more complicated than they appeared to be at first sight. He discovered that, with the parameter at high levels, the system would break apart, and the population would oscillate between two alternating values.

Ecologists had previously studied these equations. But they were looking for constants and ignored the information contained in the graphs. May and his colleagues looked at the graphs and realized the "wider implications".

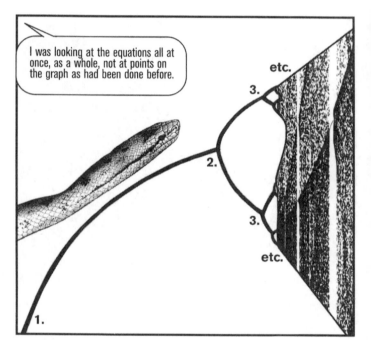

May described his discovery as the "snake in the mathematical grass" and the alterations were dubbed **bifurcations**, as we have already seen on page 18. His work confirmed the idea that biological systems are governed by nonlinear mechanisms.

Chaos in Real-Life Events

May observed that in the laboratory, animal populations do not behave chaotically. They actually fluctuate around a point according to environmental factors – they are linear. But this does not mirror what happens in the real world where populations experience the period doublings.

The answer to predicting animal populations lies with computers, where, in an imaginary world, the investigator can stipulate all the factors that could impinge on an animal's life.

Environmental randomness is introduced too, for example, by altering the number of eggs laid by an insect.

Though this is interesting, it does not totally cover all real-life events. Species interact with each other, and we can never know all the factors affecting one group of animals. Or as May says: "There is, unfortunately, no punch line to this part of the story."

Mitchell Feigenbaum: Nonlinear Patterns

Mitchell Feigenbaum, a graduate student at Massachusetts Institute of Technology, was the first to prove that chaos is not a quirk of mathematics but a universal property of nonlinear feedback systems. He provided the first significant theoretical evidence that chaos exists in many real-world situations.

During my research, I noticed that a certain pattern carried over from one non-linear system to another. It was the limit of a sequence of numbers that appeared in the calculations. On my hand calculator, this special number was 4.669.

Feigenbaum consulted his colleagues, who advised him to check his results using larger data and a computer. The computer gave a number of 4.6692016090. This convinced Feigenbaum that something was afoot.

"Imagine that a prehistoric zoologist decides that some things are heavier than other things – they have some abstract quality he calls weight – and he wants to investigate this idea scientifically. He has never actually measured weight, but he thinks he has some understanding of the idea. He looks at big snakes and little snakes, big bears and little bears, and he guesses that the weight of these animals might have some relationship to their size. He builds a scale and starts weighing snakes. To his astonishment, every snake weighs the same. To his consternation, every bear weighs the same, too. And to his further amazement, bears weigh the same as snakes. They all weigh 4.6692016090. Clearly weight is not what he supposed. The whole concept requires rethinking." (Gleick, **Chaos**, p. 174)

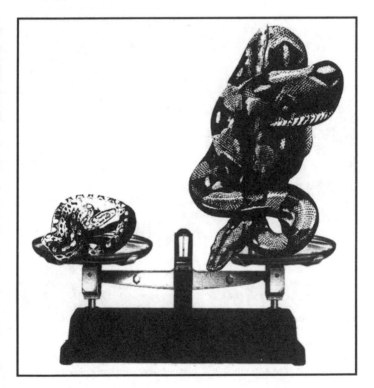

Easy Solutions to Difficult Problems

Feigenbaum had no idea why the regularity occurred. He assumed that his numerical functions expressed natural laws about systems at the point of transition between order and turbulence. Patterns in numbers implied patterns in turbulence. He eventually explained his discovery by coining the notion of **universality**. It did not explain the phenomenon – but did make a difference between beautiful mathematics and useful theory.

This was hard for physicists to swallow, for they have always believed that difficult problems require difficult solutions. So it took the scientific community some time and tribulation to accept Feigenbaum's discovery.

Ilya Prigogine: Dissipative Systems

The Belgian chemist **Ilya Prigogine** (b. 1917) is one of the true pioneers of chaos. In 1977, he won the Nobel Prize in Chemistry for his work on dissipative structures. Prigogine was the first to introduce the notions of dissipative systems and self-organization and to show that conditions which give birth to structures are "far from equilibrium".

Some parts of the universe are closed areas that operate like machines, but they form only a small part of the universe.

Most other areas are open and exchange energy or information with their environment.

Biological and social systems are open, therefore understanding them in mechanical terms will not work. Most of reality is not stable, but full of disorder and change.

Disorder to Order

Prigogine differentiates between systems "in equilibrium", "near equilibrium" and "far from equilibrium". A small population where the addition of a few births and deaths does not greatly affect the situation is in equilibrium. However, if the birth rate were suddenly to rise uncontrollably, then strange things could happen – it is far from equilibrium. In far-from-equilibrium systems, we can see matter being dramatically reorganized. There is a transformation from disorder – thermal chaos – into order. New dynamic states of matter may originate – states that reflect the interaction of a given system with its surroundings. Prigogine called these structures *dissipative structures*, because they require more energy to sustain them. In general, dissipative structures involve some damping process, like friction.

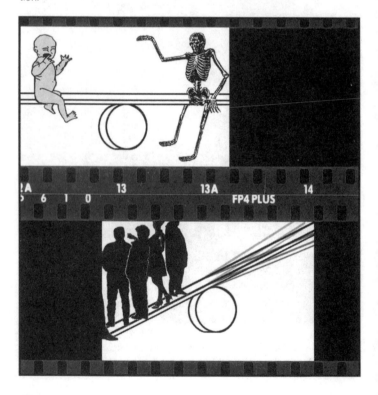

Self-Organization and Time

Moreover, when a far-from-equilibrium system enters a chaotic period, it changes into a different level of order "spontaneously" through what Prigogine called "self-organization". Initially, Prigogine's ideas on self-organization were highly controversial. He also brought **time** into the equation of chaos and complexity.

Time is what keeps everything from happening at once.

Time and the Problem of Entropy

In Newtonian physics, time was an "after thought". Newton thought that each moment was like any other. The machine can run forwards or backwards, it doesn't really matter. However, thermodynamics and its crucial Second Law placed time in a central position. The machine is running down, and time can only run one way. You cannot make up for **entropy** – the universe faces *heat death*.

Prigogine argued that time could only appear with randomness.

Only when a system behaves in a sufficiently random way may the difference between past and future, and therefore irreversibility, enter its description.

future

past

The Source of Order

In some chemical reactions, two liquids mixed together diffuse until the liquid is homogenous. They do not de-diffuse themselves. At each moment the liquid is different, and therefore "time oriented". Chemists regarded this as an anomaly.

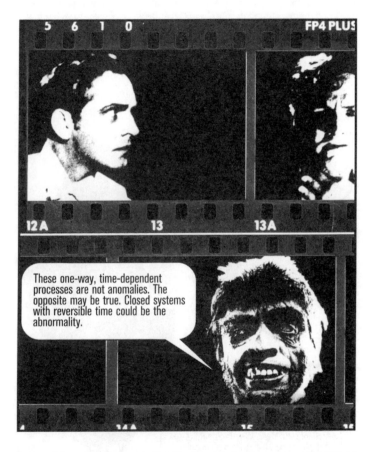

Irreversible processes are in fact the source of order – hence the title of Prigogine's most famous book, *Order out of Chaos* (1984).

Prigogine: "Far-from-equilibrium studies led me to the conviction that irreversibility has a constructive role. It makes form. It makes human beings."

Irreversible time is not an anomaly, but relates to reversible time. This is not an either/or situation. Reversibility applies in closed systems only. Irreversibility applies to the rest of the universe.

Other Features of Self-Organization

Prigogine defined self-organization as the phenomenon by which a system self-organizes its internal structure independent of external causes. Such self-organizing systems also exhibit other properties of chaos – nonlinearity, feedback, fractal structures and sensitive dependence. The French physicist **Bénard** provided a demonstration of self-organization – even before Prigogine had come up with the notion.

Bénard's experiment works as follows. He placed some liquid in a vessel and heated it from below.

At the beginning, when the temperature difference between the heated base and the cool top was low, the heat was transferred by conduction and no macro-motion was observed in the liquid. Later, however, as the temperature difference between base and top increased, a certain threshold was reached. The movement in the liquid became unstable and chaotic and then suddenly an ordered pattern appeared. The molecules of the liquid which had been moving at random suddenly exhibited a clear macro-movement in rolls which were millions of times larger than themselves. When the liquid was contained in a round vessel, the motion of the rolls formed a hexagonal pattern on the surface of the liquid. This pattern is a result of hot liquid rising through the centre of the honeycomb cells, and the cooler liquid falling along their walls. All this appears to be the result of a force, but no such force is present. The order is spontaneous. It is self-organization in action!

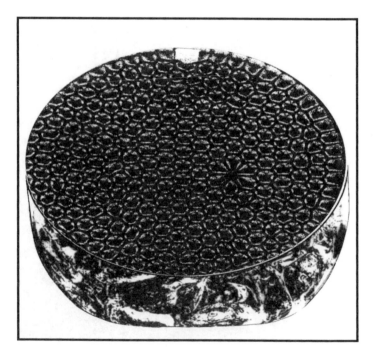

Self-organizing systems have three main features.

1. They are open and part of their environment, and yet they can attain a structure and maintain it in far-from-equilibrium conditions. This undermines the traditional view that systems must be examined as if they were isolated from their environment. These systems also run contrary to the Second Law of thermodynamics which states that they should move towards molecular disorder, and not order.

2. The flow of energy in these systems allows them spontaneously to self-organize – creating and maintaining a structure in far-from-equilibrium conditions. Such systems also create novel structures and new modes of behaviour. Self-organized systems are thus said to be "creative".

3. Self-organized systems are complex in two ways. First, their parts are so numerous that there is no way in which a causal relationship between them can be established. Second, their components are interconnected by a network of feedback loops.

Period Three Chaos

Tien Yien Li and **James Yorke**, two mathematicians working at the University of Maryland, are credited with coining the term "chaos". The term was first introduced in their much-quoted paper, published in 1975, with the peculiar title "Period Three Implies Chaos". What is *period three*?

Li and Yorke showed that it was impossible to set up a system that would repeat itself in a period three oscillation without producing chaos. Yorke explains his discovery in these words: "In *any* one dimensional system, if a regular cycle of period three ever appears, then the same system will also display regular cycles of every other length, as well as completely chaotic cycles."

Let's put this in another way. Consider a population of, say, insects. For a given population, when the *parameter rate of population growth*, **r**, is increased, the population initially increases too. Then at a critical point, two lines appear – the *bifurcation*. This corresponds to a population going from a one-year to a two-year cycle. These two lines would double again as the parameter was raised and the pattern of population repetition slowly broke down. Suddenly, chaos would appear, with whole sections of the graph blacked in.

Then, just as suddenly, windows of regularity would appear, always odd, like 3 or 7.

This meant that the population was at this moment oscillating around a 3 or 7 year cycle.

Any system that repeated itself in a period three oscillation would produce chaos. It cannot exist without it.

This technical description of chaos seems to fit the non-technical definitions of chaos. Thus, whether they intended to or not, Li and Yorke succeeded in establishing a new scientific term.

Chaos is a loaded term. Its widespread application as the name for a new science, a new perspective on the natural world, does not convey with precision or clarity the nature of the phenomena that its methodology has made apparent. Many scientists consider "chaos" to be a poor name for the new science because it implies randomness. For them, the overriding message of the theory is that simple processes in nature can produce edifices of complexity *without* real randomness.

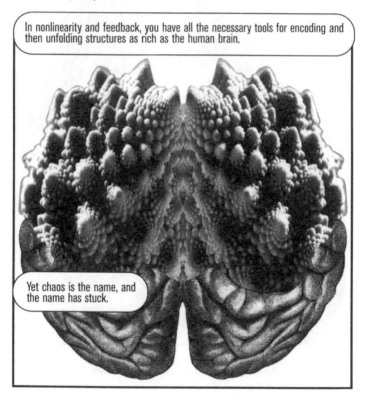

In nonlinearity and feedback, you have all the necessary tools for encoding and then unfolding structures as rich as the human brain.

Yet chaos is the name, and the name has stuck.

Towards the Edge of Chaos: Complexity Theory

In the 1980s, the study of chaos moved deeper into real-world situations. Scientists began to produce experiments which sought, and found, chaos in physical systems. This was significant because it took chaos from the realm of theoretical abstraction to being an objective characteristic of nature.

At the same time, the phenomenon at "the edge of chaos" began to attract greater attention from scientists in most disciplines. And the contours of an even newer science began to appear – **complexity.**

What is Complexity?

The nonlinear dynamic systems studied by chaos theory are complex systems in the sense that a great many independent variables are interacting with each other in a great many ways. These complex systems have the ability to balance order and chaos. This balance point – called the edge of chaos – where the system is in a kind of suspended animation between stability and total dissolution into turbulence, has many special properties.

Complexity is the new science of complex systems. It studies "life at the edge of chaos" and explores the properties of complex systems at that state.

What special features do complex systems exhibit at the edge of chaos?

The sheer richness and diversity of interactions between a host of inter-dependent variables allows complex systems to self-organize. The process of self-organization happens spontaneously – as though by magic! Think of a flock of birds taking off to fly to their place of migration. They adjust and adapt to their neighbours and unconsciously organize themselves into a patterned flock.

> Through the simple acts of buying and selling, people organize themselves into an economy.

> And it happens automatically without anyone leading the process or consciously planning it.

Atoms form chemical bonds with each other and organize themselves into complex molecules. **Spontaneous self-organization** is one of the main hallmarks of complex systems.

Adapt and Relate

The other main characteristic of complex systems is their **adaptive** nature. Complex systems are not passive – they respond actively to transform whatever happens to their advantage. Species adapt to changes in the environment. Markets respond to changing circumstances (prices, technological advances, changes in style etc.). The human brain constantly organizes and reorganizes its billions of neural connections to learn from experiences.

Complex systems also highlight the inter-relatedness of things.

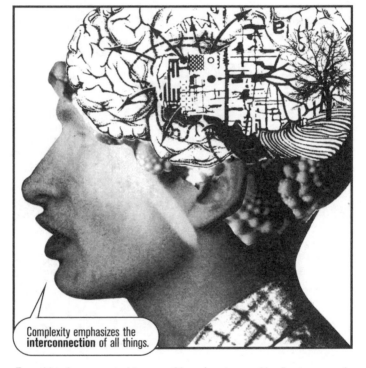

Complexity emphasizes the **interconnection** of all things.

Everything is connected to everything else: trees with climates, people with the environment, societies with each other. We no longer stand alone. Nothing does.

What is the difference between complexity and chaos theory?

Complexity is concerned with how things *happen*, whereas chaos tends to observe and study unstable and aperiodic *behaviour*. Chaos seeks to understand the underlying dynamics of a complex system. Complexity grapples with really big questions.

Science journalist **Roger Lewin** says that "as the theory of life at the edge of chaos, complexity includes the entire spectrum, from embryological development, evolution, the dynamics of ecosystems, complex societies, right up to Gaia – it's a theory of everything."

Beyond Entropy

Complexity's greatest contribution has been to show that the Second Law of thermodynamics is not the whole story. The Second Law introduces "the arrow of time" into physics and states that the entropy, or *disorder*, in the universe can only move in one direction – it can only increase. The universe is doomed to an ultimate state of absolute disorder.

Complexity shows that not **all** systems tend towards disorder or entropy.

Nature contains deep order, which is believed to "emerge" naturally.

New variables occur over time and do not require an outside force in order to "be". This is not a problem for physicists, but is more of a problem for biologists, because it appears to contradict Darwinian ideas.

A great deal of recent research on complexity has been carried out at the Santa Fe Institute, an inter-disciplinary centre of excellence which was established in 1984, with the specific purpose of developing the theory of complexity.

Together, chaos and complexity seem to be driving our world. Everything that is real is chaotic – space flight, electronic circuits, deserts, ecology of jungles, the stock market, national economies … the list is endless. And all living systems, and most physical systems, are complex systems.

Given their complementary features, it makes sense to combine chaos and complexity together.

Chaotics

This is what three European scholars, **George Anderla**, **Anthony Dunning** and **Simon Forge**, suggest: "Together, chaos and complexity spell **chaotics**."

Chaotics, they suggest, could be used to create a framework within which novel solutions to problems can be found, as new ways of thinking about and solving problems are explored.

Chaotics not only affects the answer, but the **thought** that provokes the question.

Let's see how chaos and complexity, or chaotics, applies to the physical world and how it is changing our perceptions of life, the universe and things in between.

Chaos and Cosmos

Complex dynamics are going on everywhere in the universe. Galaxies swirl around. Supernovas explode in shock waves, giving birth to stars and cauldrons of chaos. Black holes swallow up passing energy. Neutron stars spin at frantic rates. Planets display fractal patterns sign-posting chaotic processes going on at their surfaces.

Poincaré's Discovery

Before the advent of chaos theory, the solar system was seen as a perfect example of "celestial mechanics". This despite the fact that at the beginning of the 20th century the French physicist and mathematician **Henri Poincaré** (1854-1912) had shown that there were serious problems when one considered the orbits of more than two celestial bodies. He plotted the orbits of three planets qualitatively in phase space and then examined a section of their trajectories.

My solutions suggest that the presence of a third body could cause a planet to gyrate, wave or even fly off.

What Poincaré had discovered, although he and others did not know it at that time, was chaos.

Poincaré's discovery – implying that the solar system was chaotic, a few decimal points away from annihilation – was neglected for decades.

The Conditions of Stability

In the 1950s and 60s, three Russian scientists, **Andrei Kolmogorov**, **Vladimir Arnold** and **Jurgen Moser**, picked up Poincaré's work. They discovered that stability in a three-body planetary system requires two essential conditions.

The first involves **resonance**.

Quasi-Periodic Stability

For three planets to be in stable orbits, it is necessary that their resonances are not in simple ratios like 1:2 or 2:3. In order to remain stable, the planets must be *quasi-periodic* – that is, the periods never exactly repeat themselves.

If the periods do repeat themselves, the perturbation may be amplified with each orbit and resonance sets in – this is the same as positive feedback.

Small events in this situation can have a major effect.

The planets' orbits could snap, sending them whizzing off through space.

The KAM Theorem

The second condition of stability involves **gravitation**.

This condition the Russian scientists formulated as a theorem. The KAM theorem, named after the initials of their surnames, states: *If you start with a simple linear system for which a solution exists and add a small perturbation, the system will remain qualitatively the same.*

In other words, if the influence of the third planet is no bigger than the size of the gravitation attraction of a fly in Australia ...

... the three bodies will remain in stable orbits.

Unfortunately, our solar system does not rigorously satisfy these conditions.

Saturn's Moons

Studies based on the results from *Voyager II*, which flew past Saturn in 1981, have shown that many of the moons in the solar system have been in a chaotic state at some time or another, before stabilizing into quasi-periodic orbits.

Hyperion, an oblong-shaped moon that tumbles around Saturn, is in such a state at the moment.

Other moons, such as Neptune's largest moon, Triton, have cannibalized celestial satellites while in a chaotic state. Astronomers believe that the orbit of Pluto may also occupy a chaotic region.

Chaos prevents asteroids from locating at certain parts of the solar system. This is why there are gaps in the asteroid belt between Mars and Jupiter.

Gaps in orbits also exist in the rings of Saturn. These gaps appear to be the result of feedback effects of gravity, exerted by Saturn and its moons, which make the regions chaotic and therefore unoccupiable.

A Chaotic Universe

Astronomers are a long way from formulating a model of the creation of the solar system based on chaos. But we no longer see the solar system as a simple mechanical clock. It is a constantly changing, complex system.

Do you think there's a butterfly flapping its wings out there?

Quantum Chaos

The universe could itself be a product of chaos. It is commonly supposed that some fluctuations created the galaxies – fluctuations that happened near the beginning of the formation of the universe. Chaos may have played a part in this.

> The universe was in a state of near-chaos at the time, and the only valid description of what happened is a **quantum** description.

> Therefore, what happened may have been **quantum chaos**.

To understand this idea, we should look briefly at the theory of quantum physics.

Brief History of Quantum Theory

Quantum physics is a theory of the microcosm which applies only to the *atomic world*. Since the 1920s, we have known that the classical physics of Newton is only an approximation of the physics that describes the sub-atomic world.

In the mid 1800s, scientists began to realize that certain events did not correspond to Newton's laws.

For an imaginary body, known as a **black body**, the graph of radiation intensity versus frequency has a very well-known curve.

The Black Body Problem

The amount of radiation emitted peaked and then fell off. The peak would be in different places for different temperatures. No one knew what was going on until **Max Planck** (1858-1947), a German professor at the University of Berlin, realized that classical physics was not working.

At first this assumption bothered him, but it worked well, and led to a startling new discovery. Planck's constant, as it is known, was related to the structure of atoms.

Applying Planck's Constant

Ernest Rutherford (1871-1937), the British nuclear physicist, had envisaged the atomic world behaving like a small solar system, with the sun represented by the nucleus and the planets by electrons. **Niels Bohr** (1885-1962) applied Planck's constant to Rutherford's model.

I found that it explained a lot of things, such as the spectral lines of the hydrogen atom.

Spectral lines appear when the light from heated hydrogen is passed through a spectroscope. The theory predicted the position of all the lines. However, Bohr was disappointed when he applied the new ideas to the more complicated helium atom – the theory fell apart. Something had not been understood.

Probability Waves

This "something" was discovered by a French Prince, **Louis de Broglie** (1892-1987). He wondered if particles had *waves* associated with them. The type of wave he envisaged was a stationary one.

Erwin Schrödinger (1887-1961) realized that there was a need for a *wave equation*.

In 1926, **Max Born** (1882-1970) suggested that the wave function did not represent the wave itself, but only a *probability*.

This is where quantum theory now stands: probability waves in stationary humps.

Chaos in Quantum Physics

Quantum theory works in the atomic world: particles are restricted to energy levels. The lowest level is the ground level in which the system usually exists. They leave these levels when a light is shone on them (or in particle terms, when they are hit by photons), jumping to higher energy levels or excited states.

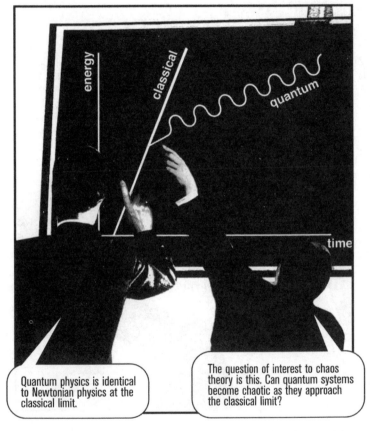

Quantum physics is identical to Newtonian physics at the classical limit.

The question of interest to chaos theory is this. Can quantum systems become chaotic as they approach the classical limit?

The question was investigated in the 1980s. And it produced a surprising result.

Physicists investigated how electrons in highly excited atoms – that is, atoms with electrons in states of extreme high energy and near to the transition between quantum and classical physics – absorb energy when radiation is shone on them.

This suppression is a subtle and delicate wave-interference effect.

Chaos In Between States

Chaos has also been investigated at quantum level by applying a magnetic field to the atom. At low levels, the electron is attracted to the nucleus, and there is no chaos.

At strong levels, the electron is so weakly attracted to the nucleus that the magnetic field overcomes it, and the electron moves around the magnetic field lines. There is no chaos.

However, in between these two states, the electron does not know where to go, and becomes chaotic.

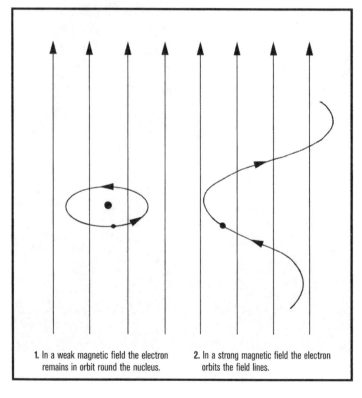

1. In a weak magnetic field the electron remains in orbit round the nucleus.

2. In a strong magnetic field the electron orbits the field lines.

Chaos is also seen when an electron is scattered by several molecules. As it moves through the molecules, its trajectory is chaotic. Small variations in its entry direction or energy make large differences to its path and where it exits. The path can only be worked out using quantum mechanics, and since it depends on initial conditions, it has chaotic characteristics.

In general, researchers have been looking for chaos in what are known as "semi-classical" systems which incorporate limited amounts of quantum effects. But the field of quantum chaos is in its embryonic stage and there is much that we need to learn.

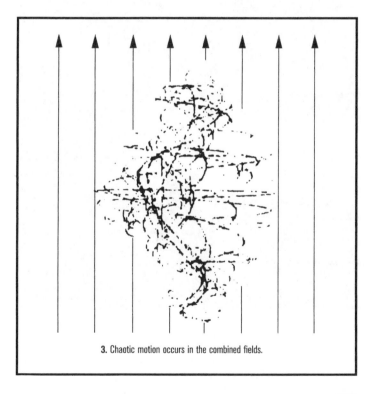

3. Chaotic motion occurs in the combined fields.

Chaos and Economics

The business environment has changed fundamentally in the decades leading up to the 21st century. The world has been linked into a single global market ruled by instantaneous transfer of capital by electronic signals. Small changes can quickly multiply in the global electronic market and lead to serious perturbations. Modern, high technology firms are radically different from traditional old-fashioned businesses. Technological innovations proliferate rapidly, making nonsense of conventional ideas of a solid lead over competition.

Value is generated in cyberspace, while jobs, pensions and welfare dissolve and become weightless. After thousands of years, this "gold standard" of monetary value is becoming obsolete. Turbulence seems to be the order of the day. Everything is "up for grabs".

Under these circumstances, chaos and complexity – or chaotics – provide us with a better understanding of what is happening than conventional economic theories. Indeed, chaos and complexity turn standard economics theories upside down – and also open up optimistic perspectives on wealth creation.

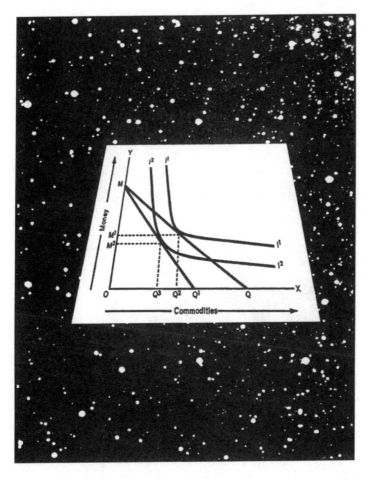

Feedback in Economics

Chaos challenges the textbook notion of economic equilibrium. This challenge comes from the concept of feedback.

Negative feedback in economic terms is analogous to *diminishing returns*; positive feedback to *increasing returns*. This way of viewing things is actually not new.

Conditions in the markets today resemble those in France in the 18th century and not those found in most economics textbooks.

Usually, it is assumed that we must wait until the final stages in order to know which way the balance will tilt in business. A firm is said to be "in equilibrium" when its net revenue is as great as it can be. This is assumed to be "its most profitable output, one that is attained through a particular and unique combination of inputs".

Therefore, under "perfect competition", there is one equilibrium point.

There is no inducement to vary the quantities of input or to change the level of output, because to move things around may affect the equilibrium point and lead to a loss of stability.

But chaos tells us there are, in fact, several equilibrium positions in this market.

The Problems with Equilibrium

Ruelle has some interesting things to say about "equilibrium".

Suppressing trade barriers has long been held to be the best way to make everybody better off. But does it?

There are never simply two countries trading, but a whole collection of connected nations and individuals. This dynamical system may *not* produce equilibrium, but chaos. Contrary to popular belief, the best laid plans of governments to design better equilibrium could in fact lead to the opposite scenario – one of total chaos.

Moreover, the idea of a single equilibrium is encouraged by the law of diminishing returns. This economic law states that as "equal increases of a variable factor, e.g. labour, are added to a constant quantity of other, supposedly fixed, factors (land, technological skills, organizational talent, etc.) the successive increases in output will after a while decline".

Chaos challenges this law, and therefore strikes at the heart of a belief in a stable economic system under competition.

George Anderla: "Orthodox economists stuck to this position basically for reasons of intellectual comfort. ... But stubbornness cannot prevail against stark reality."

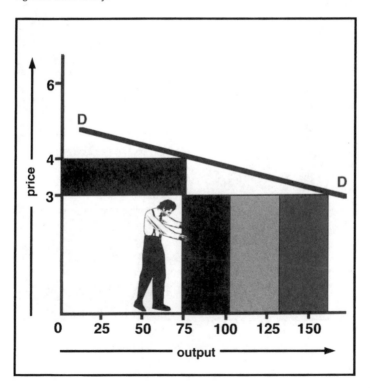

Increasing Returns in High-Tech

This one-sided "single equilibrium" interpretation of the law is now untenable, due to the rise of high-tech industries. Computers, software, optical fibres and telecommunications equipment, medical electronics and pharmaceuticals are all subject to increasing returns. This is because, from the outset, they necessitate enormous outlays on research and development, designing and redesigning, developing a prototype, and setting up tools and automated plants for manufacture.

Every time an aeroplane manufacturer, such as Boeing, develops a new aeroplane, it invests a sum in excess of half the company's net worth.

However, once the products start rolling off the production line, the cost of making additional units drops very sharply in relation to the initial investment.

How can we reconcile the conventional assumption of diminishing returns with the trend towards apparently increasing returns? **W. Brian Arthur** of Stanford University and the Santa Fe Institute has developed new insights into the crucial role of positive feedback in the economy. He realized that positive feedback makes the economy function as a *nonlinear system*.

Positive feedback drives up sales once a threshold in economy has been reached and the market has been driven to a threshold of education and promotion. As more people adopt a specific technology, the more it improves and the more attractive it looks to the designers/adopters and to would-be manufacturers and sellers.

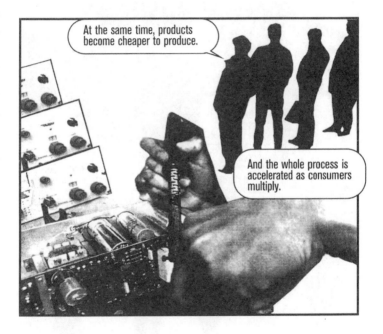

Software, once written, tested, debugged and enhanced, costs peanuts to duplicate. It can thus become a massive source of continuous, ever-increasing returns – until the producers decide that the time has come to bring out a better version.

Beware of "Initial Conditions"

Sensitivity to initial conditions can also mean life and death for a product. The best example of this is the VCR story. Sony was first on the market with Betamax, beating its rival, JVC, a small Japanese firm which had developed an alternative format, VHS. However, within a very short time, the VHS format had completely taken over the market. Traditional economics cannot explain how this happened. VHS did not divide the market as expected, but *took over*. Chaos theorists emphasize the similarities between the companies. The two video recorders came out at about the same time and had about the same price. But small "emergent-phase, chance events" tilted the competition towards VHS.

Ruelle again: "Although a system may have sensitive dependence on initial conditions, this does not mean that everything is *un*predictable about it. Finding what *is* in fact predictable and what *isn't*, is a deep, unsolved problem."

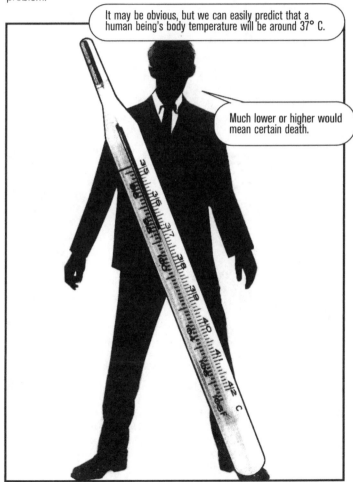

The End of Neo-Classical Economics

If you take all the factors together, the production function shows increasing returns. So if you double all the factors, output more than doubles. Firms are able to make bigger profits by "simultaneously raising their output, lowering costs and reducing prices as a means of fostering the sales". The neo-classical assumption of "perfect competition" is thus thrown overboard.

How to Play Monopoly

More companies are now working under increasing returns, and this situation has led to a *de facto* **monopoly.** The success of Microsoft is based on the fact that once the initial cost of developing particular software such as Windows '95 is recouped, the returns continue on an ever-increasing spiral – leading to a monopoly.

Simon Forge: "Is this like a Japanese game of Go – the more you win, the easier it is to encircle your foe?"

Chaotic Management

Modern "scientific management" first came into vogue with the publication of **Frederick W. Taylor**'s book *The Principles of Scientific Management* (1911). Taylor (1856-1915) was an American industrial engineer who originated scientific management in business. He was obsessed with the need for maximum efficiency. Over the last thirty years, however, the notion of scientific management has changed – especially with the emergence of computers. The Harvard Business School introduced the concept of **Strategic Planning** in the 1960s and 70s.

This stressed the need to integrate standard business functions such as production, accounting and marketing with a systematic approach to an overall strategy.

But experience showed that highly mechanistic plans and mathematical forecasts do not always work.

Massachusetts Institute of Technology then introduced **system dynamics**.

But both of these management techniques are risky, and are based on subjective assumptions and value judgements.

According to Simon Forge, this approach is like "driving using the rearview mirror" – trying to judge the road ahead by what goes on behind.

Anticipating Future Breakthroughs

So, how can management anticipate with some degree of confidence that a technological and industrial breakthrough is about to emerge?

Technological breakthroughs sometimes occur by chance – small events, considered unimportant at the time, trigger a chain reaction leading to a new technological discovery.

The discovery of penicillin is one example of a breakthrough that came about as a result of a random event in medical research.

Other breakthroughs, however, are the result of years of research. The moon landing and the events that followed it are a good example of this.

This type of breakthrough is more common nowadays. Research is multi-disciplined and on a very large scale.

According to George Anderla, integrating the dynamic concept of increasing returns with the philosophy of large-scale multi-disciplinary research – the holistic approach – produces "creeping breakthroughs".

Creeping breakthroughs are best studied with the Butterfly effect of chaos theory.

Enablement and Forecasting

But anticipating "creeping breakthroughs" is not enough. Management has to think not just about what a new technology can do, but also to consider its back-up system. To be productive, inventions must have co-inventions, as things rarely stand alone. Long-range bombers, for example, were a good idea (except for those who were to be bombed), but until a way could be found to refuel them in the air, the idea could not be implemented.

The concept of simultaneously thinking about inventions and co-inventions is known as "enablement".

Creeping breakthroughs and the requirements of enablement mean that old methods of forecasting technological development are no longer valid. As the early bird sets the standard, to miss out at the initial stages is to miss the chance to call the tune. A systematic means to spot breakthroughs is therefore of paramount importance.

The conventional approach is to assess the relative importance of the different elements of a possible breakthrough and identify any enablers still outstanding. To recognize the potency of novel ideas, and then synthesize the new usage scenarios and concepts, perhaps by combining old concepts in new ways.

Forecasting has to be a holistic and continuous process with feedback, sensitive dependence and nonlinear developments kept firmly in view.

Chaos and Cities

Cities have changed; and so has our view of cities. Cities have changed from ordered, controllable entities to untamed and untameable environments. Our image of cities has changed from the positivist, humanist and Marxist-structuralist city of modernity to the ever-changing, chaotic city of postmodernism.

There is a pluralistic kaleidoscope of cultures and sub-cultures in each city: from Asian, Italian and Chinese, to "straights" and gays, productive and desolate regions, pedestrian walkways and "no-go" areas. Nothing is stable, nothing is true, and nothing matters for very long.

Cities are microcosms and mirrors of societies and cultures at large. Thus to develop a thorough understanding of cities we need to take account of most, if not all, of the diversity that produces a contemporary city. The conventional ideas of cities as "architecture-writ-large" cannot be easily related to the theory of cities as social, cultural, economic and institutional systems.

Social systems are not easy to relate to spatial form. So our current understanding is overwhelmed by their complexity and diversity.

This is where chaos comes in. Chaos provides us with deeper insights into spatial order in the city.

All cities have some irregularity in most of their parts and as such are ideal candidates for application of fractal geometry. In fact, cities have quite distinct fractal structures in that their functions are self-similar across many orders and scales. The idea of neighbourhoods, districts and sectors inside the cities, the concept of different orders of transport nets, and the ordering of cities in the central place hierarchy which mirrors the economic dependence of the local on the global and vice versa, all provide examples of fractal structures.

Fractal properties of cities enable geographers and urban planners to study population densities, land use and spatial texture which reflect spatial juxtapositions.

Fractal Cities

Fractal geometry can be applied to cities in at least two ways. First, in terms of visualizing urban form through computer models and computer graphics; second, through measurements of patterns in real cities and their dynamic simulation.

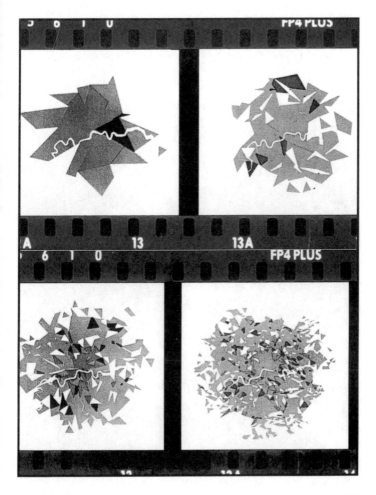

Michael Batty, Professor of Spatial Analysis and Planning at the University of London, is a pioneer in the field of "fractal cities".

Batty: "Using fractal geometry, we can explore the geometry of cities first by fixing size and varying scale, and then by fixing scale and varying size. This idea is of central importance to the development of a theory of a fractal city."

His work has shown that the key relations of fractal dimensions for the city involve relating population and its density to linear size and area. These relations are structured in incremental or cumulative form.

This interesting fractal is actually London (population density of).

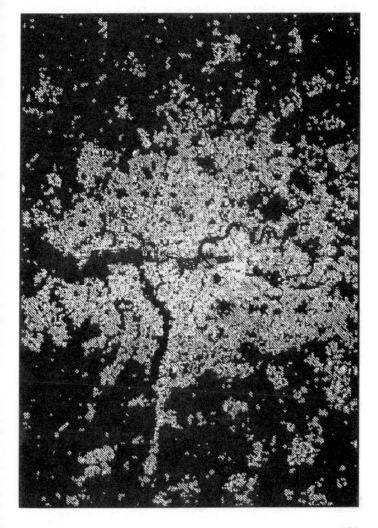

Fractal Skylines

Batty: "These fractal relations appear to have greater rationale than those used traditionally, and the whole approach shows how careful one must be in defining and measuring densities. One conclusion of my work is that much of the work on urban density theory and its application over the last forty years will have to be reworked in the light of these developments."

Skylines of cities – such as Manhattan – can also be fractal in nature.

Dissipative Cities

Apart from fractal cities, recent urban analysis has revealed a number of other types of chaotic cities.

Dissipative cities are the products of Prigogine's theory of dissipative structures and its application. The theory of dissipative cities has been developed, amongst others, by **Peter Allen**, Professor at the International Ecotechnology Research Centre of Cranfield University. His work involved building computer models of infrastructures of localities in a region, each with residents and jobs. The individuals migrate to get employment, and employers offer or remove jobs depending on the market. This migration between localities and the introduction and extraction of economic activities, create a local "carrying capacity" and this leads to nonlinearities and feedback loops for the system which links population and manufacturing activities.

This in turn leads to an evolutionary process by which new urban centres grow and others decline.

The interplay between interaction and fluctuations, on one hand, and dissipation on the other, brings about a new landscape. Allen later applied this model to Brussels.

Local and Global Chaos

Prigogine's ideas on self-organization have also led to the notion of "self-organized cities" or "chaotic cities". In cities, self-organization occurs in two forms: local or *microscopic* chaos and global macroscopic or *deterministic* chaos. Local chaos is a result of the behaviour of the individual components of a city, e.g. the movement of cars on a motorway.

Deterministic chaos arises when, as a consequence of self-organization, the individual parts are attracted by a few attractors. The city jumps back and forth chaotically from one attractor to another. For example, on a freeway, the movement of cars is randomly distributed at nights and almost uniformly distributed at rush hours.

There is thus a shift from chaos to order, and then back to chaos.

The play between chaos and order shows up in daily routines and not just in long-term development.

Control or Participation

Chaos has brought a new perspective to our understanding of cities as urban spaces. It has shown that factors that control the evolution of a city are self-organizing systems and as such are themselves uncontrollable.

Batty: "From this perspective follows a new type of action in the city, a new way of urban planning, which aims not to control but to participate."

Chaotic Architecture

It is hardly surprising that fractal shapes are used in postmodern architecture. Architect **Bruce Goff**, for example, was amongst the first to use strange attractors to organize a force-field of movement inside some of his houses.

In her competition-winning design for Cardiff Bay Opera House, **Zaha Hadid** used fractal geometry to create a building that uses the language of planes that enfold difference in continuity. However, the design became controversial, as it was too postmodern for the taste of many, and her conception was never realized.

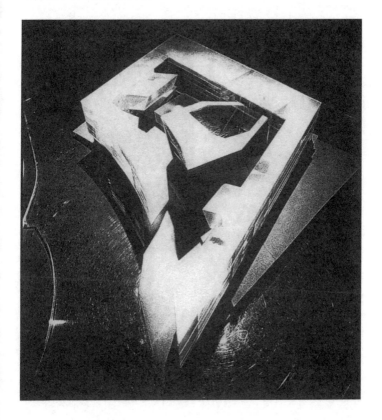

Such notions of chaos as nonlinearity, period doubling and feedback are becoming increasingly common in postmodern architecture. These ideas, in the words of **Charles Jencks**, architect and guru of the postmodern architecture of chaos and complexity, generate "an architecture of waves and twists, an architecture that undulates and grows and diminishes continuously *and* abruptly".

But the use of chaos is not limited to postmodern architecture. Some traditional buildings express the same ideas. Fractal scaling can be seen, for example, in the baroque Paris Opera Building, designed by **Charles Garnier** (1825-98) and built between 1861 and 1875. It consists of an elaborate combination of styles based on an underlying harmony. A walk down Rue de l'Opéra reveals the self-similar details of the building: the closer you get, the more detail comes into view.

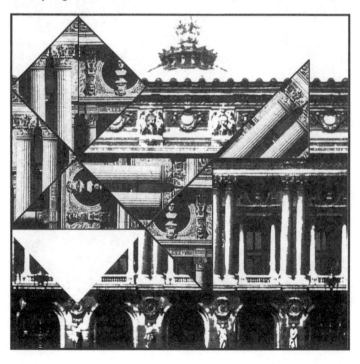

Chaos and the Body

The conventional model depicts the human body as a machine. The heart beats like clockwork, the nervous system is a telephone exchange, and the skeleton is just so many joints and hinges.

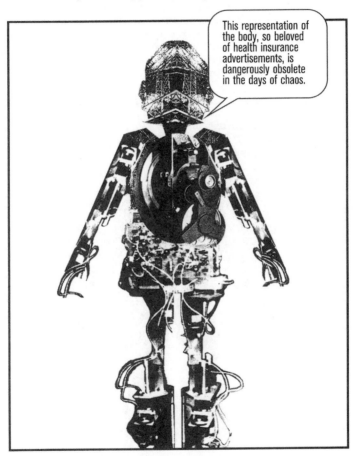

Biologists, physiologists and medical experts are now beginning to portray human physiology as a holistic system full of fractals and chaos.

Body Fractals

Our body is simply covered with fractals – they are everywhere from the circulatory system to the lymph system, the lungs, the muscle tissue, the calyx filters in the kidney, the small intestine to the folding patterns on the surface of the brain. These fractals make the body flexible and robust. Because they are self-similar, parts of the body's fractal structures can be injured or lost without serious consequences. Fractal structures also increase the surface area available for the collection, distribution, absorption and excretion of a host of important vital fluids, as well as toxins, that regularly pass through the body.

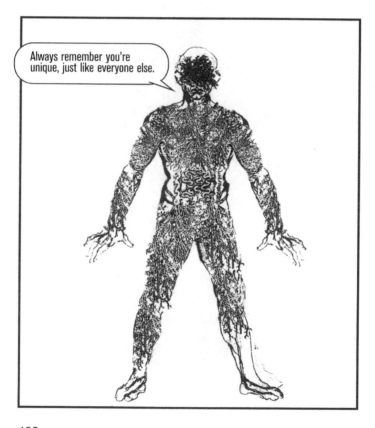

The Heart's Attractor

Chaotic dynamics is also present in the body. It is the product of feedback that is constantly going on between numerous parts of the body.

When the ECG (electrocardiograms) of the heart are plotted in phase space, they reveal a "spider-like" strange attractor.

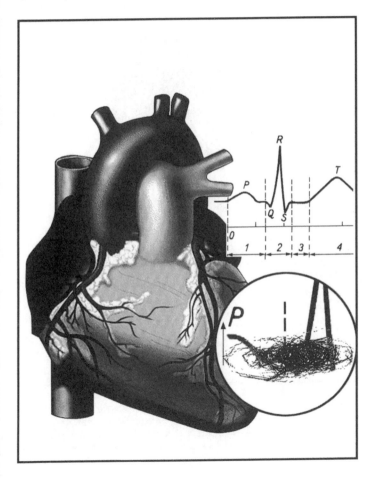

Chaos in the Heart

Period doubling provides a clue to the onset of a heart attack. In a healthy heart, electrical impulses move smoothly through the muscle fibres that force the heart's ventricle to contract and pump blood. In their contracted states, the muscle fibres become impervious to electrical signals. This period is called *refractory time*.

Researchers have discovered that when a group of heart muscle fibres have a refractory time that is longer than the interval between heartbeats, turbulence could set in.

This knowledge has been used to develop a prototype for an intelligent heart pacemaker. This device constantly monitors the heart, recognizes when undesirable chaos is setting in, detects what will happen in the next split second, and sends an electrical signal to the heart to stop it going wrong.

Chaos and Good Health

But not all chaos in the body is bad. There is natural background chaos in the body – for example, in the brain activity – which performs useful functions. Loss of this chaos can lead to abnormal functions. The seizure in epilepsy, for example, may appear as an attack of chaos, but it is in fact due to *loss* of chaos. It is the product of an *abnormally periodic order* within the brain.

Conventional wisdom holds that ill health is a product of stress and other factors that disturb the body's normal periodic rhythms.

Chaos tells us that irregularity and unpredictability are important characteristics of good health.

Chaos and the Brain

One of the discoveries of chaos theory is that the brain is organized by chaos.

The human brain is a complex nonlinear feedback system. It contains billions of neurons, connected to each other.

Signals in the brain travel in endless feedback loops, carrying vast amounts of information.

Although we know that certain regions of the brain perform certain functions, activity in one area can trigger more neuronal responses throughout a large region.

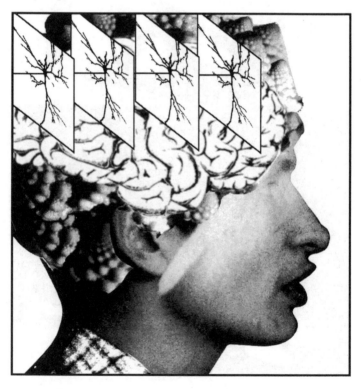

Experiments have shown that the brain has strange attractors – indeed, countless strange attractors, one each for a particular activity. Plots of EEG (electroencephalogram) activity in the brain show one particular type of strange attractor when a person is at rest, but quite another attractor when the same person is solving a mathematical problem. A healthy brain maintains a low level of chaos which often self-organizes into a simpler order when presented with a familiar stimulus.

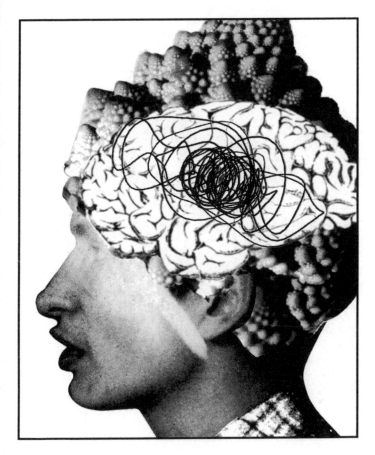

A Chaos Model of Consciousness

If we assume that the states within the brain are connected with consciousness, we can arrive at a radical model of consciousness. So how does chaos theory help us to understand consciousness?

Basically, neurons only fire a signal when they are activated by incoming signals from other neurons. The concept of phase space is used to picture what is happening inside the brain. Each neuron is considered to represent one variable. So, in phase space, each neuron is given one dimension. There are therefore 10 billion dimensions. If consciousness is, in fact, related to the activity of these neurons, then through this model, we have a representation of consciousness that can be analyzed.

What conclusions can we draw from this point?

First, its path is chaotic. The system may be deterministic, but the behaviour of the point is unpredictable. From this it can be said that we can never truly predict how people will behave.

Second, the movement of the point, whilst being chaotic, is *not* random. It follows a strange attractor. The strange attractor could be the phenomenon that we know as "personality".

Third, this model is not algorithmic – it is not predictable or sequential. It is fluid and flexible.

Fourth, there is no limit on the number of states that this system can reach. The number of neurons is finite, but the points in phase space are limitless. Thus consciousness itself is limitless.

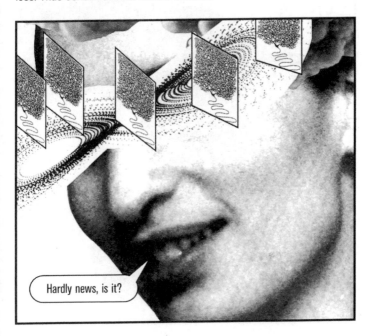

Hardly news, is it?

Chaos and Weather

Chaos theory owes a great debt to the weather – without it, it might not have developed in the way that it did. The weather is in fact a quintessential chaotic system.

Not surprisingly, it displays a fractal structure – thus displaying self-similarity. What we see on the planetary scale, we see generally when we zoom down to a continental scale and to a national scale. And all of the weather's components, from temperature, air pressure, wind speed to humidity, are sensitive to initial conditions. As it constantly folds back on itself – its *iteration* – the weather displays a vast range of chaotic behaviour on many scales. But it remains within the broad range of a strange attractor we call **climate**.

Chaos notwithstanding, we will continue to try to predict the weather from observations of certain initial conditions. Today's weather prediction models have about a million variables and are developing all the time.

But not surprisingly, meteorologists do not always get their predictions right.

Long-Term Weather Prediction

But what about long-term prediction? What will the weather be like in the next century? Long-term predictions are totally different from trying to discover what the weather will do tomorrow or next week. We are not looking for an individual trajectory in the attractor in this case, but the shape of the whole climatic attractor itself.

Global climate is subject to feedback. There is always the danger that positive feedback could accelerate even the minutest perturbation by human beings into an environmental catastrophe. However, negative feedback keeps the atmospheric temperature stable. Given the infinite number of positive and negative feedback loops, it is impossible to say what real fate awaits us.

149

What About the Greenhouse Effect?

However, repetition in weather patterns may mean that a trajectory is simply circling one of the butterfly wings – it can do this once, twice or a thousand times. There is no pre-ordained number of times that it can do this. Therefore we ought to be careful about making predictions about the "greenhouse effect", for example. A series of warm winters and hot summers may simply mean that the system is revolving around one part of the phase space. It does not necessarily mean that long-term permanent change has set in.

Chaos and Nature

Chaos and complexity reflect a new sensibility about the world around us. Not so long ago, people believed that science would simply conquer all ignorance. With technology, we would simply dominate the natural world. Chaos tells us that nature is not a simple system, ready to bend to our will. Indeed, nature can strike back, and does – as when we breed resistant strains of micro-organisms by the widespread use of antibiotics.

Scientific Safety

Until quite recently, people associated science with two goals – knowledge and power. It would eliminate superstition, ignorance, disease and poverty. But we are now becoming aware of the price to be paid for this simplistic view of nature. The achievements of science that were made in terms of this view were great, but one-sided.

Both from chaos theory and from our experience of the environment, we have a new understanding. When even deterministic systems cannot be predicted, uncertainty becomes a major concern.

So now there is a third great goal for science: **safety**.

We can even think of man-made risks as a sort of "chaotic complexity". For the complex natural systems of matter, energy and life which had gone through their cycles for aeons have now been perturbed. New substances, and new forms of energy, have been injected into the perennial natural processes.

How many "butterflies" have been created whose consequences are unknowable?

We know now that the natural world around us can no longer be guaranteed to run smoothly and safely for our benefit. New diseases, global pollution, species extinction, and climate change are all the results of the unexpected impacts of our simple-minded science and technology on nature.

The New Nature

So long as the image of science was of simple deterministic equations, these new phenomena would have been difficult to conceive. But with chaos, we can think again about nature in relation to ourselves. Formerly nature was "wild", and with our science-based technology we "tamed" her. The regular laws of her behaviour were revealed, and she was put under the yoke of our machines. But now, in the age of chaos, we can recognize a new state of nature: a presence that is "feral".

We can imagine these states of nature in terms of polluted systems that have escaped from our domestication. But they are not just wild, or "natural" as in the previous pre-chaos system. Rather, as we see with goats, rats or rabbits introduced into a new habitat, there is a destructive, perhaps catastrophic imbalance among the species.

Is It Safe?

Chaos and complexity provide us with the conceptual tools for coping with these new problems. We know that science is incapable of producing firm predictions of the future states of such chaotic complex systems. In particular, it is impossible for science to prove whether something is perfectly "safe". Whether we accept, or tolerate, some particular risk, will depend only partly on what the scientific experts tell us. It will also depend on the judgements, and the value-commitments, of all those affected by the problem.

When safety is the issue, rather than knowledge or power, conventional science is an invaluable servant for decision-making, but it can be a very misleading master.

The new chaos-based understanding of nature requires a new notion of the appropriate form of scientific practice. This new practice of science is called "post-normal" science.

Post-Normal Science

Post-normal science is the brainchild of two philosophers of science, **Silvio Fontowicz** and **Jerry Ravetz**.

Ravetz: *In pre-chaos days, it was assumed that values were irrelevant to scientific inference, and that all uncertainties could be tamed. That was the "normal science" in which almost all research, engineering and monitoring was done. Of course, there was always a special class of "professional consultants" who used science, but who confronted special uncertainties and value-choices in their work. Such would be senior surgeons and engineers, for whom every case was unique, and whose skill was crucial for the welfare (or even lives) of their clients.*

Fontowicz: *But in a world dominated by chaos, we are far removed from the securities of traditional practice. In many important cases, we do not know, and we cannot know, what will happen, or whether our system is safe.*

We confront issues where facts are uncertain, values in dispute, stakes high and decisions urgent. The only way forward is to recognize that this is where we are at. In the relevant sciences, the style of discourse can no longer be demonstration, as from empirical data to true conclusions. Rather, it must be dialogue, recognizing uncertainty, value-commitments, and a plurality of legitimate perspectives. These are the basis for post-normal science.

Post-normal science can be illustrated with a simple diagram.

Close to the zero-point is the old-fashioned safe "applied science". In the intermediate band is the "professional consultancy" of the surgeon and engineer. But further out, where the issues of safety and science are chaotic and complex, we are in the realm of "post-normal science". That is where the leading scientific challenges of the future will be met.

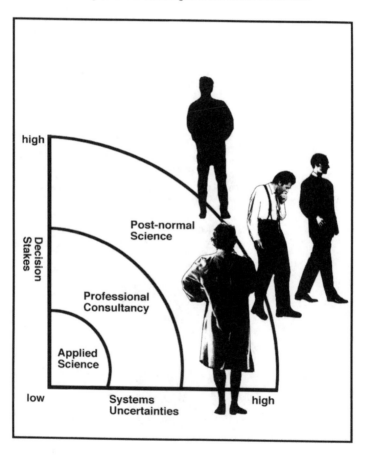

Post-normal science has the following main characteristics.

Fontowicz: *In post-normal science, Quality replaces Truth as the organizing principle.*

Ravetz: *In the heuristic phase space of post-normal science, no particular partial view can encompass the whole. The task now is no longer one of accredited experts discovering "true facts" for the determination of "good policies". Post-normal science accepts the legitimacy of different perspectives and value-commitment from all those stakeholders around the table on a policy issue. Among those in the dialogue, there will be people with formal accreditation as scientists or experts. They are essential to the process, for their special experience is used in the quality control process as the input. The housewife, the patient, and the investigative journalist, can assess the quality of the scientific results in the context of real-life situation.*

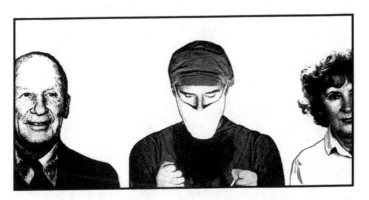

Fontowicz: *We call these people an "extended peer community". And they bring "extended facts", including their own personal experience, surveys, and scientific information that otherwise might not have been in the public domain.*

Post-normal science does not replace good quality traditional science and technology. It reiterates, or feedbacks, their products in an integrating social process. In this way, the scientific system will become a useful input to novel forms of policy-making and governance.

Chaos and the Non-West

Chaos theory and complexity are tools for understanding. But these new sciences contain understanding that has been indigenous to non-Western societies.

Indeed, it is how non-Westerners have traditionally seen themselves, their environment, their place in the cosmos and what they have traditionally done.

Natives in India, for example, have been using fractals as an art form for centuries. The Indian craftsmen can draw this famous Kolam pattern very quickly – and it can be found in their *durrees* and other types of floor coverings (see bottom of this page).

Symmetrical fractals can be seen adorning the ceilings of most medieval mosques – here is a vestibule ceiling from the Chenar Bagh Madresseh (school) in Ishfahan (below).

Islamic art and design has always used simple, fractal patterns to generate complexity as a mental tool to focus the intellect on the contemplation of the Infinite.

But more than that: the insights of chaos and complexity can be found in most non-Western cultures. Humility before nature, richness and diversity of life, generation of complexity from simplicity, the need to understand the whole to understand a part – these are the things that the non-West has not only believed in but acted upon. They are intrinsic in most non-Western worldviews.

Traditional non-Western agricultural techniques, from the use of underground aquifers in the Middle East to pest control with birds in Sri Lanka, have been shown to be ecologically more sound than modern agriculture.

In non-Western mystical systems, such as Buddhism and Sufism, self-looping contradictory statements are used to take the minds of students to the edge of chaos and then to enlightenment through self-organization. A student asks …

Thus a movement is set up where the mind's understanding of truth and falsehood continually fold back on each other.

Much of the alternative development literature, from the critiques of the Latin American schools of dependencies, the Indian criticism of modernization to the Muslim scholarship on Westernization, has argued that sensitive dependence on initial conditions would not allow the Western model of development to work in their region! Over and over again, the critique of non-Western experience has urged that the complex initial conditions of non-Western civilizations and environments have been insufficiently understood, that valued elements in the non-Western holistic context have not been taken into account, and thus the grandly devised deterministic programmes could not achieve their projected ends. There is a litany of case studies to substantiate this point.

Twenty years on, chaos encapsulates the same criticism in mathematics and eye-catching computer graphics. One could say that the arrival of chaos substantiates that critique as authoritative.

It is hardly surprising that complexity has often been compared to Taoism.

Brian Arthur, Stanford University Professor and a former director of the Santa Fe Institute, says: *The complex approach is total Taoist. In Taoism there is no inherent order. "The world starts with one, and the one become two, and the two become many, and the many led to myriad things." The universe in Taoism is perceived as vast, amorphous, and ever changing. You can never nail it down. The elements always stay the same, yet they are always arranging themselves. So it's like a kaleidoscope: the world is a matter of patterns that change, that partly repeat, but never quite repeat, that are always new and different.*

Just as there is no duality between man and nature in non-Western world-views – such as the Islamic, Chinese and Hindu – so there is no duality in complexity.

Arthur: *We are part of nature ourselves. We're in the middle of it. There is no division between **doer** and the **done-to** because we are all part of this interlocking network.*

Finally, Arthur admits: *Basically what I am saying is not at all new to Eastern philosophy. It's never seen the world as anything else but a complex system. But it's a world view that, decade by decade, is becoming more important in the West – both in science and in the culture at large. What is happening is that we are beginning to lose our innocence, our naiveté.*

It seems that after centuries of denigrating non-Western ideas and notions, science is coming back to non-Western viewpoints.

Criticism of Chaos

For the last few decades, the quest for the Truth that can be undisputedly proved has accelerated, partly due to the breakdown of all belief systems in the West and partly due to the awesome power for mathematical manipulation that the computer has unleashed. In mathematics, this quest has manifested itself in a number of fashions and fads. Each fashion was supposed to provide us with new all-encompassing insights into nature and reality and bring us face to face with ultimate reality. In the 1950s, "game theory" was supposed to describe human behaviour and thus allow us to control and manage it. In the 1960s, René Thom's "catastrophe theory", which describes the dynamics of certain nonlinear systems, was projected as a universal law that explained everything from embryological development to social revolution. Then came "fuzzy sets", for which equally grandiose claims were made. Now we have chaos and complexity.

Are chaos and complexity simply a new fad? Can we expect chaos to be there in the next century, or will it be supplemented by another fashion?

Peter Allen has consistently argued that chaos is not a discipline in its own right. Rather, it's just a subcomponent of nonlinear dynamics which is itself just part of complex systems. "In reality, the important aspect is the origins and evolution of structure and organization in complex systems – not the trivial occurrence of sensitivity in strange attractors. However, chaos may be used in nature to provide 'noise' with which to maintain adaptability and surprise."

Ian Stewart, Professor of Mathematics at the University of Warwick and one of the leading authorities on chaos in Britain, says: "The term 'chaos' has escaped its original bounds, and in so doing has to some extent become devalued. To many people, it is no more than a new and trendy term for 'random'. Take some system with no obvious pattern, declare it to be an example of chaos, and suddenly it is living on the intellectual frontiers instead of being boring old statistics again. Chaos has become a metaphor, but far too often the *wrong* metaphor. Not only is the metaphor being extended to areas where there is no reason to expect a dynamical system, but the very implications of the metaphor are being misrepresented. Chaos is used as an excuse for the absence of order or control, rather than as a technique for establishing the existence of hidden order, or a method for controlling a system that at first sight seems uncontrollable."

This is not surprising. Such abuses emerge whenever a deep intellectual concept becomes fashionable.

Stewart: "The same happened to Einstein's relativity theory which was widely used in the United States as an excuse for social inequality. 'Everything is relative, as Einstein says', became the chant. Not so. The most interesting thing that Einstein said is that some things, notably the speed of light, are *not* relative."

It is not just that chaos does not offer ready-made solutions to everything, but it is also "difficult to reconcile a complex universe with the presumed simplicity of its rules".

Stewart: "Many of the great mysteries of science are emergent phenomena. Mind, consciousness, biological forms, social structures – it is tempting to leap to the conclusion that chaos and complexity hold the answers to these mysteries. However, at least as currently conceived, they do not and cannot. The role of chaos and complexity has been crucial and positive: they have caused us to start asking sensible questions and to stop making naïve assumptions about the source of complexity or pattern. But they represent only a tiny first step along a difficult path, and we should not let ourselves be carried away by overambitious speculations based on too simple a notion of complexity."

The danger is that chaos and complexity would become the "bible" – a new theory of everything. The overzealous champions are already projecting the new science as some sort of universal calculator.

The real importance of chaos is its capacity as a new tool for solving problems and a new way of thinking about nature, the physical world and ourselves. In this respect, it is a field with great potential and could truly shape our future.

Further Reading

The most popular and engaging account of *Chaos* (Sphere, London 1988) is by James Gelick, who also gave the new science the cachet of a pop style. In *Complexity* (Dent, London 1993), Roger Lewin tried to outdo Gelick. But Mitchell Waldrop kept his sense of balance in *Complexity* (Viking, London 1992).

There are numerous coffee-table accounts of chaos, all with stunning pictures and great graphics. The best amongst these are John Briggs, *Fractals – The Patterns of Chaos* (Simon & Schuster, New York 1992) and Michael Field and Martin Colubitsky, *Symmetry in Chaos* (Oxford University Press, Oxford 1995).

More mathematically inclined explorations of chaos can be found in John Briggs and David Peat, *Turbulent Mirror* (Harper & Row, New York 1989) and Nina Hall (ed.), *The New Scientist Guide to Chaos* (Penguin, London 1992).

Even deeper experiences are provided by B.B. Mandelbrot, *The Fractal Geometry of Nature* (W.H. Freeman, San Francisco 1982), Stephen H. Kellert, *In the Wake of Chaos* (University of Chicago Press, 1993), Edward Lorenz, *The Essence of Chaos* (UCL Press, London 1995), David Ruelle, *Chance and Chaos* (Penguin, London 1993) and Stuart Kauffman, *The Origins of Order* (Oxford University Press, Oxford 1993).

Michael Batty and Paul Longley's exploration of chaotic cities is considered to be ground-breaking: *Fractal Chaos* (Academic Press, London 1994). In *Chaotics* (Adamantine Press, Twickenham 1997), George Anderla, Anthony Dunning and Simon Forge give an insightful account of new economics and management theories of chaos and complexity. Charles Jencks does an excellent PR job for chaotic architecture in *The Architecture of the Jumping Universe* (Academy Editions, London 1993), and Barry Parker provides a remarkable, but mathematically drenched, tour of *Chaos in the Cosmos* (Plenum Press, London 1996). Ilya Prigogine and Isabelle Stengers' *Order Out of Chaos* (Fontana, London 1985) is one of the first detailed explorations of chaos – a classic by any account! In *Uncertainty and Quality in Science for Policy* (Kluwer Academic, Dordrecht 1990), Silvio Funtowicz and Jerome Ravetz explore the management of risks in chaotic times.

Critical perspectives on chaos are provided by Ian Stewart, *Does God Play Dice?* (Basil Blackwell, Oxford 1990), Jack Cohen and Ian Stewart, *The Collapse of Chaos* (Viking, London 1994) and the special issue of the prestigious journal, *Futures* (vol. 26, no. 6, July/August 1994), *Complexity: Fad or Future?*, edited by Ziauddin Sardar and Jerome R. Ravetz.

Index

Biographies

Ziauddin Sardar has always lived on the edge of chaos. Starting off as a science journalist, he has spontaneously self-organized himself into a television reporter, a futurist, a scholar of Islamic studies, a cultural critic and a visiting professor of science policy. This chaotic behaviour naturally forced him towards a strange attractor called scribbling; and periodic doubling led to the production of over two dozen scholarly and popular tomes. He is the author of the classic studies, *The Future of Muslim Civilisation* and *Islamic Futures*. His more recent books include *Mathematics, Muhammad, Cultural Studies* and *Postmodernism* (co-authored) in the *Introducing* series, *Postmodernism and the Other, Barbaric Others* (co-authored), *Cyberfutures* (co-edited) and *The Consumption of Kuala Lumpur*. Sensitive dependence on initial conditions required a family life: he is married with three children and lives in London.

Iwona Abrams – illustrator and printmaker.
Graduated from the Krakow Academy of Fine Arts in Poland and the Royal College of Art in London.
Her work has been exhibited in Britain and abroad.
Commissions: Vintage, Women's Press, Heinemann International, Cambridge University Press, *The Sunday Times*, *GQ*, *The Economist*, *The Observer*, Spero Communications.

Acknowledgements

Thanks are due to Gail Boxwell for her chaotic assistance.

Iwona Abrams thanks Teresa Frodyma for her role as Cordiallia Cauliflower.

Typesetting by Iwona Abrams.